Airport Baggage Handling Systems

The competition between airports demands higher-quality services to satisfy passengers. A Baggage Handling System (BHS) serves functions such as baggage sorting, screening, and storage. A successful BHS means bags move between areas as travellers do. Handlers load/unload bags and transfer them between the airport's areas. Automation can save money and bring safety and operational benefits. Warehouses as well as the automotive industry are more advanced technologically in comparison to the airport and especially the baggage handling business. The concept of the Baggage Factory (BF) is an approach (based on Industry 4.0) that simplifies the processes from the moment passengers drop off their bags at check-in till the destination (chutes, carousels, and so forth).

Airport Baggage Handling Systems: Using the Baggage Factory Approach to Support AI Optimisation, Decisions, and Design Processes introduces the features of the BF concept and presents how BHS designers can use AI technology to tackle many BHS problems and concerns. The book bridges the gap between airport BHS designers and experts in AI and optimisation. It describes in detail the field of baggage handling using algorithms for sorting bags or optimising the flow. The way the systems are designed is discussed, and a behind-the-scenes look at the BHS industry and how it affects the daily lives of travellers is presented.

International and multidisciplinary in approach, this book is an ideal resource for practitioners, students, and researchers involved in the air transportation industry, Tourism, Systems Engineering, Layout and Design, Artificial Intelligence, Assembly Automation, and Logistics fields.

Dr Brahim Rekiek earned his license in Physics from the Université Abdelmalek Essaadi, Morocco, in 1993, his D.E.S. in Production Engineering and Robotics in 1996, and a Ph.D. degree in artificial intelligence in 2000 from the Université Libre de Bruxelles. He has worked for more than 20 years at Fabricom Airports Systems and Beumer in the main sectors of airport BHS, and Automotive and Logistics. He has experience in a variety of projects related to the design and analysis of baggage handling systems for airports worldwide, including tasks such as layout, capacity analysis, and simulation. Dr Brahim has published many scientific papers on the subject of optimisation, design, and layout and has supervised many scholars and industrial research projects in the last 20 years as well as published one book.

Airport Baggage Handling Systems

Using the Baggage Factory Approach to Support AI Optimisation, Decisions, and Design Processes

Brahim Rekiek

Université Libre de Bruxelles

CRC Press
Taylor & Francis Group
Boca Raton   London   New York

CRC Press is an imprint of the
Taylor & Francis Group, an **informa** business

Designed cover image: Shutterstock

First edition published 2024
by CRC Press
6000 Broken Sound Parkway NW, Suite 300, Boca Raton, FL 33487-2742

and by CRC Press
4 Park Square, Milton Park, Abingdon, Oxon, OX14 4RN

CRC Press is an imprint of Taylor & Francis Group, LLC

© 2024 Brahim Rekiek

ISBN: 978-1-032-55926-1 (hbk)
ISBN: 978-1-032-55925-4 (pbk)
ISBN: 978-1-003-43292-0 (ebk)

DOI: 10.1201/9781003432920

Typeset in Nimbus font
by KnowledgeWorks Global Ltd.

Publisher's note: This book has been prepared from camera-ready copy provided by the authors.

This book is dedicated to my parents.
To my sisters and brothers
To my wife, Aaida, to my children, Saad, Inas, and Sara.
To my beloved country.

Contents

PART I   Introduction

PART II   *From Intra-Logistics to Airport BHS*

PART IV   *Challenges*

Foreword

Despite the high number of companies and people working in the field of baggage handling systems, there is a lack of books describing the inner behaviour of this business.

It is also very difficult to find theoretical background about algorithms for sorting bags or optimising their flows. In fact, there are not many books about the BHS (as the only topic) on the market, while there are many about airport management in general. Thus, the first purpose here is to transmit the knowledge accumulated in projects during the last decades to the young generations of engineers, academicians, and so forth.

It is known that it is always a challenge that new engineers share when they join, for the first-time, BHS companies such as Beumer, Daifuku, Vanderlande, Siemens, TGW, and Fabricom Airport Systems. This new book entitled *Airport Baggage Handling Systems: Using the Baggage Factory Approach to Support AI Optimisation, Decisions, and Design Processes* by Dr Brahim Rekiek is an important contribution in the domain which tries to fill this void.

The book also presents some new ideas for airports of the future. It tries to bridge the gap between airport BHS designers and experts in Artificial Intelligence and optimisation. The purpose is to give AI practitioners access to the BHS challenges and, at the same time, to make it easy for BHS designers to tackle the BHS problems using AI. It is particularly interesting to see how AI and statistics can be applied to operational planning for the whole of the baggage handling system to avoid operational peaks, reduce operation costs, minimise passenger waiting times, reduce busy periods, and more. The book gives a short introduction to the main AI techniques, and proposes some references to the most well-known publications in the field. The author has expertise in the optimisation field and knows how to design systems that get the most of the available resources.

It introduces the basics of the Baggage Factory concept. The idea of changing the materials handling of bags into a solution like that is used often in the intra-logistics, especially in distribution warehouses. The baggage handling system should adopt a pull rather than push concept. The batch building concept using bag stores is certainly a more economical method to handle baggage. There are a couple of innovative ideas and a detailed description of many of the processes involved in airport baggage handling systems.

The author has published many scientific papers in the subject of optimisation, design, and layout. In this book, he shares his 15 years of rich experience in the field of airport baggage handling systems design (and intra-logistics systems in general). Dr Rekiek has been working on many logistics systems and developed many optimisation techniques.

The book can provide newcomers and experienced BHS professionals with a broad inventory of existing technologies and ways forward to improve the performance and the efficiency of that unknown underground area for most people, and that is the BHS of the airport.

Eng. Alain Chalon
Systems and BHS Expert
March 2023

Preface

A great idea can be obtained of just how amazing airports are when considering a large airport in Belgium (namely Brussels Airport[1]) in which over 20 million people flow through each year. This becomes a startling point when considering that the population of Belgium is around 11 million only. The air industry is a crucial part for achieving economic growth and development through facilitating integration into the global economy and providing vital connectivity on all levels (e.g., local, national, regional, and international levels). In 2019, the number of passengers world-wide reached almost 4 billion persons. There exists a high competition between airports, and this demands higher-quality services to satisfy passengers' requirements. This forces airports to use efficiently and effectively their existing infrastructure and continuously upgrade it.

The Baggage Handling System (BHS), which means handling of bags and containers of bags, is a daily business of every airport and is one of the major indicators of an airport's level of service. The BHS handles bags from check-in desks to departure laterals, and those coming out of aircraft via arrival inputs to arrival reclaims or from transfer to departure laterals. The 45-minute "connection time" is one of the selling arguments of all major airports. A typical BHS is composed of several parts that can deal with complex functions such as baggage sorting, screening, and storage. The measure of a successful BHS is quite simple. Do bags move from one point to another as fast as travellers can? An inefficient BHS leads to passenger and airline dissatisfaction. The most important performance indicators are the average cost per handled bag, the number of mishandled/damaged/delayed bags, the number of employees needed, the customer satisfaction, etc.

Handlers are responsible for loading/unloading bags and transferring them between the different areas of an airport. Their job entails repeated lifting, pulling, pushing, twisting, and stretching their arms and back, and, consequently, their injuries might cost a lot of money to airports. Therefore, using robots can spare human beings from lifting dozens of tons per day, as the robots are capable of handling any type of bags, weighing up to 50 kg. In general, the investment in robots can save insurance costs and bring added safety and operational benefits, while fitting in a smaller space.

The variability in handling rates of bags by human operators can sometimes become a bottleneck in airport operations, especially considering the fluctuations that occur across the days, months, and seasons. While fully automated systems can be a solution, their high investment costs and the possible negative social impact make them not acceptable to many airports. Thus, airports may find it challenging to decide whether to replace their current workforce with robots. A more acceptable and efficient solution could be to adopt a semi-automated system that retains the human

---

[1]This was the value of 2022, just after the recent COVID-19 pandemic.

handlers at the right processes while redesigning the work organisation to find efficient solutions.

Humanity has come a long way before arriving where we are today. First came steam and then the first machines that mechanised some of the works our ancestors did. Next, machines powered by electricity and combustion engines allowed mass production and created an economic value. In the late twentieth century, the third era of industry started with the emergence of computers and automation, when machines began to replace humans in the production plants. Thus, many industries, especially warehouses and automotive, have entered the era of Industry 4.0, with robotic machines connected to IT systems based on machine learning algorithms that are able to learn and to control the robots without input from the operators.

The BHS and the automatic warehouses seem to be completely different, but when you look closely, then the similarities begin to stand out. Both systems receive, sort, store, and forward items (bags or parcels) on to their destination. Warehouses are, in general, pull systems; they are triggered when an order is submitted resulting in its items being picked (pulled) from storage, collected into individual orders, and then batched by location into loads ready to be placed on a vehicle to take them to their destination. Warehouses, as well as the automotive industry, were and are still more advanced from a technological point of view in comparison to the airport, especially the baggage handling business. Therefore, the airport business must be inspired by the other industries!

The Bag Factory is an approach (from Industry 4.0) that drives up efficiency by simplifying processes from the moment passengers drop off their bags at the check-in till their destination. Like in a factory, end-to-end solutions are fully needed to master the process. Thus, we need to look at the processes, their integration, and not only the technology. Drop-off, reclaim, automated guided vehicles, baggage storage using cranes, on-demand bags, and batch build are some features that the Baggage Factory has taken from the automotive and warehouse industries, but there is still a long way before they are generalised in airports.

Nowadays, many passengers check-in online and use the drop-off machine to hand over the bags to the sorting system. The baggage drop-off machines deliver a wide range of benefits, including shorter queues, reduced labour costs, and PAX travel time saving. These drop-off points can also be installed remotely, at hotels, at the airport's parking, and so forth. Nevertheless, the problem is still the same underground (nobody sees it) when it comes to operators to load heavy bags into containers. Ergonomics can bring some solutions to these issues.

At the arrivals, on-demand baggage reclaim virtually eliminates the risk of a bag being reclaimed by the wrong passenger. In the future, rather than collecting their bags from the carousel immediately after arrival, passengers will be able to decide when they want to collect their bags. This not only reduces the pressure on passengers but also allows them to spend time in a reclaim shopping area, which provides the airport with additional revenues. The approach can also be used for the shopped goods that were brought on and bought before the departure flight and stored in storage until the passenger returns home.

The Baggage Storage System is foreseen to efficiently manage bags that enter the BHS before their flight has been allocated to a sorting destination. The conventional approach is to store bags on the floor. The high bays (cranes, shuttles, etc.) originated from warehouses are used more and more and can hold a thousand of bags. Their strong advantage is the utilisation of the height of the building, allowing to store more bags per ground surface, combined with the random storage principle allowing to individually access each storage position, which leads to an efficient use of the storage space combined with fast storage and extraction average times.

The shuttle (bus) self-driving vehicles in the airport move passengers from the car parking lot to the terminals. Similarly, there are vehicles that transport autonomously the containers from one location to another and interface directly with the conveyors at each of the unload/offload locations. Instead of circulating on conveyors and carousels, the bags can be transported at high speed on autonomous vehicles. These self-directed vehicles with embedded intelligence circulate on tracks (DCV, AGV, and Totes).

Nowadays, most BHSs are push systems; passengers check-in or drop-off their bags and these are then pushed towards their destinations. This approach demands a lot of space and resources and suffers from inefficiency. The automated batch build is a process to load bags into containers in batches. These batches are scheduled and managed automatically. The batch build concept is intended as a fact to execute the right job at the right time, at the right place, where the required resources are available, as efficiently as possible, within a statistically optimised lower-cost batch window. Thus, the aim is to transform a BHS from a push system to a pull system. The aircraft will be the instigator in the system by requesting its next load and the aim is to deliver all the items destined for any aircraft as efficiently as possible. Early bags need to be held in baggage storage until they can be handled with the rest of the on-time bags being processed for the flight.

The impact of the recent pandemic on air travel has been dramatic, making it the worst aviation crisis ever. This event has completely changed the known paradigms; new ones should be developed. As the pressure to cope with changes in passenger volumes continues (high service level quality, short period of time to build flights, etc.), airports must be prepared to embrace innovative technology to ensure maximum efficiency. The batch build concept, along with the utilisation of robots and AGVs, the new AI modules, and other advanced technologies, offers a promising alternative to conventional approaches such as the push concept and manual handling. By leveraging these solutions, we can enhance our ability to achieve the desired targets and objectives.

With the era of Big Data, the need reappeared to solve optimisation of complex problems of unprecedented large sizes. Machine learning and computational biology are examples of the application domains where it is possible to formulate these problems with a large number of variables. Classical optimisation algorithms are not designed to deal with instances of this size; new approaches are needed. Thus, the challenge is to come up with more advanced optimisation algorithms capable of working in the Big Data setting.

Only a few problems can be solved by structured conventional optimisation methods (e.g., linear programming, branch and bound, etc.) alone. On the contrary, the Artificial Intelligence (AI) techniques are limited in their search for the optimal (best) solution. The combination of AI and conventional optimisation techniques can provide problem-specific reasoning and powerful numerical optimisation tools. By combining AI and Machine Learning (ML), more effective optimisation procedures that fit the BHS can be obtained.

In the future, airports will be able to offer passengers the option of a 100% self-service journey, where human interaction is eliminated due to the use of automated, self-service technology at every single step of the journey. Staff will be able to focus on providing excellent customer service without the need to perform basic operational processes[2]. Automation and efficiency will go together, with the technology allowing air travellers to walk through every process at the airport, from the door of their taxi to their seats on the plane. We will get to the level where the processes no longer dominate the airport journey, and the passengers have more freedom to do what they want at the airport, whether working, eating, shopping, or just boarding the plane. Full automation!

The loss of low-paying human jobs is always a concern whenever automation is introduced. Thus, maintaining the integrity of the processes with less operators could become an obstacle. Additionally, the stakeholders and investors are, in general, reluctant to invest actively in new technologies.

Nevertheless, the benefits of the Baggage Factory (automation) model could outweigh the concerns for many airports. In dangerous working environments, the health and safety of operators could be improved dramatically using automation. The processes could be easier to control when there is data (IoT) at every device leading to reliable productivity, so the results could be an increase in profits, efficiency, etc. The question is not if the Baggage Factory will be adopted or not, but how quickly it will be in use. As with big data, Artificial Intelligence and Machine Learning, the early adopters will be rewarded for their courage jumping into new technologies, a new environment, etc.

---

[2]Low-cost airlines are likely to use technology to maintain their low level of service while reducing staff costs.

Acknowledgements

I would like to express my sincere gratitude to the many individuals and organisations who contributed to this work.

Firstly, I would like to thank the whole team at the former Fabricom Airport Systems and, in particular, Alain Chalon, Jean Michel Delcourt, Stéphane Corteel, Arnaud Marchand, José Dumont, and Pierre Delplace for their helpful insights and contributions.

I would also like to extend my thanks to the team at Ultra Electronics Tisys, and especially to Jean-Claude and Fabien Betend.

Special thanks go to the team at Beumer/Crisplant, and particularly Dirk Fransen and Moureaux Noel for their invaluable support.

Many thanks to the Université Libre de Bruxelles, especially to Prof. Emanuel Falkenauer. I would also like to express my gratitude to Prof. Alain Delchambre for his unwavering support throughout my academic career.

Lastly, I would like to thank Dr H.A. Saleh, Ahmed Rekiek, and all my family members for their ongoing encouragement and support.

Dr B. Rekiek

Author

Dr Brahim Rekiek holds a license in Physics from the Université Abdelmalek Essaadi, Morocco, obtained in 1993. Additionally, he earned a D.E.S in "Production Engineering and Robotics" in 1996 from the prestigious Université Libre de Bruxelles. Building upon his educational accomplishments, he successfully earned his Ph.D. degree in artificial intelligence from the same university in 2000.

With over 20 years of experience, he has made significant contributions to the field while working at renowned companies such as Fabricom Airports Systems and Beumer, among others. Throughout his career, he has focused on various sectors including airport BHS, Automotive, and Logistics. He has played a role in designing and analysing baggage handling systems for airports worldwide, undertaking tasks such as layouting, capacity analysis, simulation, software development, research and development, and more.

He has numerous publications in the field of optimisation, design, and layouting, with many scientific papers and book chapters to his name. Additionally, he has supervised numerous scholars and industrial research projects over the past two decades. His expertise and dedication to sharing knowledge led to the publication of the book *Assembly Line Design* (Springer-2006), which has been recognized as an invaluable resource. Furthermore, he has served as a reviewer for various international journals.

With a solid educational background, extensive industry experience, and notable achievements in scientific research and project management, he is a seasoned professional at the intersection of IT software, artificial intelligence, and engineering.

List of Figures

List of Tables

Part I

Introduction

1 Introduction

1.1 INTRODUCTION

You can get an idea of just how amazing airports are when considering a typical large airport in Belgium in which over 20 millions people can flow through per year. Consider that the population of Belgium is only 11 millions or so; that's a pretty startling statistic! According to IATA[1], this number will increase in the next decades, despite the pandemics negative effects.

International airports host several types of visitors (e.g., passengers, passenger's family, staff, etc.), and most of them are passengers. Normally, large airports can handle more than a hundred of flights every day (one flight every 2 minutes, almost 24 hours a day!), serving hundreds of thousands of people. That means million(s) of local, national, and international passengers pass through the airport each year and that is a lot of people. Each passenger carries one up to three bags, and that is a lot of bags too.

Air traffic has increased during the last 25 years, and according to SITA [164], the number of passengers has expanded by an average of 3–5% per year since 2012. In 2018, the number of passengers worldwide reached 4.3 billion a year[2]. This forces airports to use their existing infrastructure as efficiently as possible. At the same time, the tremendous competition between airports asks for a higher-quality services to satisfy the different customers. Passengers' expectations are clear: they want to check in their baggage as quickly as possible, then, on arrival, quickly collect their baggage undamaged[3].

The baggage handling is the daily business of every airport, and it is one of the major tasks at airports and a crucial indicator of the level of services. Inefficient baggage handling leads to non-satisfied passengers and airlines [40, 127]. Therefore, most airports try to deliver the highest level of services to their customers. Well-organised baggage handling with short baggage's transfer times from one flight to another (transfer baggage), for departure flights (outbound baggage) or for arrival flights (inbound baggage) attracts new airlines and motivates passengers to use the airport [25, 55, 192]. Thus, efficient usage of existing baggage handling helps to avoid infrastructure expansions which are quite costly, and efficient baggage handling system also leads to labour cost savings.

The performance of the Baggage Handling Systems (BHS) is measured by the required time that a bag can be transported from one point to another within the

---

[1]IATA (International Air Transport Association) international aviation organisation with more than 250 member airlines comprising more than 90% of all scheduled international air traffic.

[2]See IATA web site: www.iata.org/passenger-forecast

[3]Checked-in baggage disappears behind a wall and, if everything goes as planned, they reappear on an arrival carousel at the destination airport.

DOI: 10.1201/9781003432920-1

system. The bags lost rate is also crucial to guarantee a certain level of service for the passengers. For many people, checking bags at the airport is still a fear. Bags are lost daily[4], and baggage handlers tend to manhandle the bags. It is amazing how bags are badly handled at BHS facilities; even carton boxes at factories are more gently handled than bags. Nobody is complaining! Concerning lost bags, customers can be less tolerant of mishaps.

Increase in baggage volumes is leading airports to invest in new and efficient baggage handling technologies in order to maintain revenues. Indeed, their operations and maintenance (O&M) management is asked more than ever to reduce costs and lower the risk of system breakdown [111, 156]. You are probably wondering why airports have not come up with a better system by now. Simply, it is due to the service level to passengers, the Return On Investment (ROI), daily interactions with the unions, neighbourhoods and disruption, etc. It is not that airports have not looked for solutions, and it is a matter of policies, decisions, priorities, etc.

In 1990, decisions made by the Denver airport authorities affected their baggage system project [39]. Indeed, when construction started on the new Denver airport, it was supposed to come with a brand-new automated BHS (local and transfer). The goal was to replace the standard reliance on manual labour with a fully automated BHS. This is not what exactly happened; it was a disaster. Doing things quickly and forcing the changes are never a good idea!

Airport services can be classified as air-side and land-side. At the air-side, airlines need space for aircrafts, runways, facilities for routine maintenance, areas for passengers and crew lounges, etc. Air-freight companies need specific areas for loading and unloading cargo aircrafts, storage, etc. At the land-side, to meet passengers' needs, airports must be accessible by public transportation plus have parking space. They must have areas for ticketing, displays, check-in, shops, accommodations, lounges, restaurants, etc.

The BHS is the interface between the land-side and air-side of the airport. It is a system that transports baggage items from check-in desks to laterals, transfer infeed, buffers, and loading station. The system handles bags coming out of aircraft into arrival reclaims or from transfer inputs to departure laterals and from check-in counters to departure laterals, etc.

A BHS is composed of different technologies which are used to transport bags from one end to another, scanners to scan the labels, sorting machines and junctions to route bags to their destination, etc. [91]. Thus, the BHS is one of the key functions of any airport, and as many support systems, it is a necessary evil! Nobody likes it, but everybody needs it. The BHS of an airport plays a crucial role in keeping travellers happy. It also makes a difference in an airport's ability to attract and keep the airlines satisfied by the offered services [23, 41, 81, 197].

---

[4]According to SITA-Baggage Report 2017 [164], 5.73 bags per thousand passengers got lost in 2016 – a 70% reduction over the past 10 years. It is a drop compared to the previous years; the rate of mishandled baggage should drop even further.

1.2 DESIGN PROCESS

As stated earlier, the BHS is the poor element of the airport. In general, when the design of the BHS starts, it is almost all the other facilities are done. Thus, the BHS designers have to deal with the remaining space and try to squeeze the equipment's as much as possible. The BHS must be fitted within the size constraints (such as shops, parking, etc.) of existing buildings, in worst scenario, or already globally designed in the majority of the cases.

The term *design* is the process of specifying an object (product, program, etc.) that satisfies a collection of *constraints* which mean something that is either satisfied or not. It is a characteristic of many design problems that new constraints emerge as *decisions* are made [15, 58]. As the human design is time consuming, many attempts have been made to investigate the use of semi-automatic design methods.

People have the impression that the design is a "cut-and-paste from old design" activities. This is not always the case, as the *creativity* has a major role in design. Design is an iterative approach, where first, a design is created and optimised, which is then analysed, experimented or tested in use, to determine its quality, etc. The design complexity is not due to the physical, material, or procedural factors; rather, it depends on understanding a problem and making well-founded decisions.

There are some general design steps that the designer has to follow: (1) formulating the problem to be solved, (2) breaking it down into sub-problems, (3) grouping ideas that must be discussed and challenged, (4) evaluating and redesigning (if needed) the current design, and finally (5) implementing the proposed solution. Table 1.1 shows some basic values of the design philosophy [133, 145].

Many computer-aided design (CAD) software, various computer simulation, and analysis and optimisation tools are used by airport's designers. The design of efficient airports is a problem of considerable importance. Many airports have adopted the Concurrent Engineering (CE) approach to improve the development of their products and services. The main aim of CE is to integrate product and process development

Table 1.1

Values of the design

Value	Target
Total Cost	to reduce
Availability	to maximise
Reliability	to optimise
Performance	to maximise
Energy efficiency	to improve
Design	to be open
Project Time-line	to be short
Expandability	to be easy

in order to reduce the design lead-time, to improve its quality and cost, etc. Once the data is defined, the layout is made, then the simulation is used to check the operation and hence, reduce the system cost and to improve its service quality (see Figure 1.1).

The elaboration of the *logical layout* of a system consists of distributing the departments, zones, functions, etc., while the *physical layout* deals with the disposition of these elements, resources, conveyors, buffers, etc. in the building (tacking into account the available space). The objective is to minimise the total cost by simultaneously integrating the design (*e.g.* space, shape, cost, *etc.*), the operation constraints (*e.g.* time, precedence constraints, availability, *etc.*) and the designer desires (*e.g.* tasks complexity, *etc.*). Improvements can result from an enhancement of the process as well as from a well-designed system. A well-designed system means an architecture in which two areas that are characterised by a big flow are placed as close as possible. There are many possible solutions to place items on a given surface. This gives rise to the well-known 2D/3D placement problem (2D/3D Bin Packing) [117].

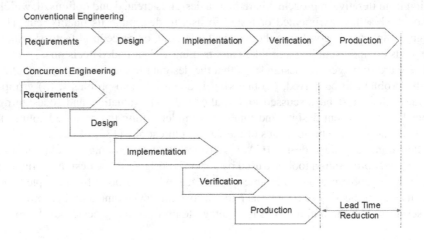

Figure 1.1 Concurrent engineering approach to system design

The design or redesign of an existing airport (or part of it) is a long and expensive process, and the decision-making process is a complex task. Hence, if one decides to undertake the airport (re)design task, a careful attention should be given to testing the various parameters and scenarios before embarking on any implementation. Different factors can affect the performance of the airfield at any given time. This can range from the available number f check-in desks, the number of chutes, etc. The ability to identify and quantify any eventual benefits to the operation of the BHS in any of these areas is vital for any airport.

It is clear that the most salient problem facing the world today is how to effectively mitigate the impacts of natural and man-made hazards and disasters. No place is isolated from the rest of the world when it comes to these impacts. Every place is exposed to all types of these hazards and disasters, and mainly to geophysical hazards, climatological hazards, and technological and man-made hazards. Designing for disaster, before it happens should be a must.

While the people involved with handling disasters may not be "designers", they are still making conscious design decisions. People think of designers as artists, but they are not just there to make things visually appealing; their mission is also to create more efficient and effective products and processes. There is surely room for designers in disasters, as there is room for designers in every aspect of life.

1.3 BAGGAGE HANDLING SYSTEM AUTOMATION

The handlers can sometimes become a "bottleneck" in the BHS due to their variable rates of handling bags, especially in some crucial areas like loading and unloading. This can lead to operational inefficiencies, especially in an environment that puts a lot of pressure on the handlers capacity due to growing passenger numbers. To go for automation, a robot can be a practical solution for the problem at hand. Indeed, robots can reduce the manual workload risks and speed-up the processes.

The higher level of accuracy due to automation can enable members to reduce mishandling rates by tracking each bag at all steps of its processing. By speeding up reconciliation and flight readiness, passengers' satisfaction levels will also be enhanced. This makes the viability of a purely manual operations, with its inherent inefficiencies, increasingly harder to justify. In the case of semi-automated systems, the operator controls the loading device manually while in fully automated stations, a robot does the job.

The loading rate[5] depends on the used baggage loading solution, which can be manual, semi- or fully automated (see Figure 1.2). Nowadays, many airports still do not use semi- or fully automatic systems, and this is mainly due to space limitation, high investment costs and low flexibility. And also because automation may not always be efficient [55, 124]. Thus, semi-automation, keeping the handlers in the system (at the right position and processes), and redesigning the work organisation (by integrating more automation) can be more acceptable and hence efficient.

The loss of low-paying human jobs is always a concern when automation is introduced. Thus, maintaining the integrity of the processes with less operators could become an obstacle. Indeed, avoiding technical problems that could cause expensive outages is always a concern. Additionally, the stakeholders and investors are in general reluctant to invest actively in new technologies[6] [39].

1.4 INDUSTRY 4.0

First, came steam and the basic machines that mechanised some of the work our ancestors did. Next, was the discovery of machines powered by electricity and internal

---

[5]The loading rate of a working station is defined as the number of bags loaded into containers per a time interval.

[6]Denver airport experience in 1990 to install new fully automated systems was not a good example !

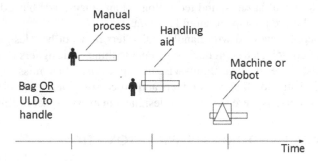

Figure 1.2 From manual processes towards automation

Table 1.2
Features of Industry 4.0 system

Feature	Description
Interoperability	Machines, devices, and staff are connected and communicate with each another.
Information transparency	The system creates a virtual copy of the installation (data collected via sensors). Also called the "digital twin".
Technical assistance	The systems are able to support humans in making decisions and solving problems and to assist humans with tasks that are unsafe for humans.
Decentralised decision-making	The ability of virtual systems to make basic decisions and become as autonomous as possible.

---

combustion engines which allowed mass production (assembly lines) and created an economic value, which marked the second revolution. In the late 20th century, the third era of industry came up with the emergence of computers and automation, when robots and machines began to replace humans on production. Advanced production systems which include electronics, information technology, and robotics were the outcome of the information age on manufacturing [94].

Now, we have just entered the era of Industry 4.0, in which computers and automation come together in a new way, with robotics connected to computers equipped with *machine learning algorithms* that are able to learn and to control the robots with very little input from human operators (that was a kind of science fiction a couple of years ago!). Table 1.2 gives the main features of Industry 4.0 system.

The merging of the virtual and the real worlds through virtual-physical systems and the resulting fusion of the technical and the business processes are leading the way to a new industrial age (defined by the "smart factory"). In the smart factory, cyber-physical systems monitor the processes of the factory and make decentralised decisions. The physical systems become Internet of Things (IoT), cooperating with each other and with the operators in real time. These smart factory resources and processes; provide significant real-time quality, time, resource, and cost advantages in comparison with conventional systems [116].

Today, billions of machines, systems, and sensors worldwide are communicating with each other and share divers information. Thus, in the future, this will not only enable companies (including airports) to make their processes more efficient, it will give them greater flexibility when it comes to change their systems/processes to meet the market requirements [50]. One scenario is that the machines will organise themselves, supply chains will automatically assemble themselves, and the orders will be transformed directly into the production processes. Yet people will remain essential in the Industry 4.0 world – as the creative thinkers will use their intelligence to develop all the processes and procedures in advance and will write software to convey that information from/to the machines. That is what is called the self-organising factories (or swarm). Thus, the virtual and real production worlds are merging [60, 188].

Whether they involve digital methods (virtual reality), lightweight robots, or 3D printing, new technologies for Industry 4.0 are already a reality. In this changing world, the airport business should adapt itself. The question is not if Industry 4.0 is coming or not but how quickly it will. As with big data and other business trends, the early adopters will be rewarded for using this new technology, new production environment, etc. and those who avoid change risk becoming irrelevant and left behind the others, as usual [116]. And, as Einstein A. said, "Insanity is doing the same thing over-and-over again and expecting different results". The airport's business will surely follow!

1.5 ERGONOMICS

At the beginning of the aviation, the baggage handlers were waiting for the passengers at the apron to collect their bags and to put them in the aircraft cargo hold. Later-on the passengers and bags flows were separated in the terminal building at the check-in. Passengers dropped off their bags at check-in counters and the BHS transports the bags to the aircrafts. With this separation, airports collect the bags beforehand and load them into aircraft cargo holds. Thus, the baggage handling became predictable, enabling airports to guarantee a certain loading time for the aircrafts. Handlers function at airports was thus born.

Thus, we moved from a situation where every passenger (suppose a flight with 100 passengers, 150 bags in total) transports himself his bags, to a situation where mechanical systems and a couple of handlers are carrying the 150 bags (3 tonnes if each bag is 20 kg). Quite a challenge! Imagine this process for many flights, day long. Those handlers are responsible for loading / unloading / storing bags from inbound/outbound aircraft flights and transferring them between the different areas of the airport. They work in all types of weather, all over the airport, etc. In general, the handler's job entails to repeatedly lift, pull, push, squat, twist, kneel, and stretch of the arms and back. This makes the handler's job one of the more challenging material handling jobs.

Baggage handling has often been overlooked at many airports, despite passengers' needs. This applies to many airports worldwide, which started with a single runway, one terminal with hundred employees, to multiple terminals and runways with several tens of thousands of employees. One of the main topics is the musculoskeletal injuries amongst workers, accidents that resulted in a wide range of injuries. Annually, baggage handlers injuries cost a lot of money to airports.

Following the many ergonomic studies that have been conducted in several airports, many technologies, new processes and mechanical handling aid solutions (e.g., loaders, unloaders, lifters, etc.) have been introduced in a couple of airports [98]. Figure 1.3 illustrate the evolution of technologies and processes within the baggage handling business, from conveyors and check-in systems to EDS, robots, etc.

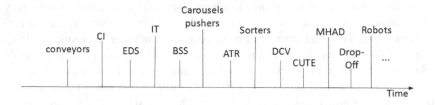

Figure 1.3   Evolution of baggage handling systems

Applying ergonomics to any task can help the operators to be safer, more efficient and more comfortable. However, so far, there are still lots of things to do with aim to improve! Indeed, from the perspective of the stakeholders, replacing labours by robots is more than welcome, from the other side it is always a massacre. Thus, it worth to never under-estimate or to over-estimate the phenomenon, just give it right value. The proof, is that often, airports unveil a series of measures intended to improve the health and well-being of their employees.

When designing a BHS, the balance between the trio (cost, ergonomics and throughput) should be foreseen in order to design a viable solution (see Figure 1.4).

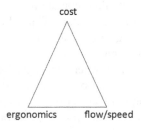

Figure 1.4   Automation and ergonomics!

Thus, Ergonomics is about making life easier! –for operators!

1.6 OPTIMISATION AND ARTIFICIAL INTELLIGENCE

Optimisation can be found everywhere; it is a part of human nature to strive for goals and to optimise the actions towards these goals. Problems can be represented with models and formulated in the language of mathematics. Mathematics and physics describe the world around us in terms of physical concepts and theoretical objects. A mathematical model is always a simplified representation of a true physical object that is being investigated.

In general, optimisation problems are expressed as the minimisation of a cost function that depends on a set of input parameters. The optimisation of the processes of a business is essential for keeping its competitiveness. As computers, solvers and techniques improve, some complex problems can be solved to optimality.

The Combinatorial Optimisation Problems (COPs) are characterised by a finite number of feasible solutions. Although the optimal solution of such problems can be found by an enumeration, especially for practical problems. We can observe a tendency to use *heuristics* rather than exact methods, as the enumerative methods ask for a lot of computing time for big, complex problems. A *metric* is needed to identify a successful search (*i.e.* indicate if the *goal* looking for was reached or not). This metric could be *binary* ("found" and "not found yet") or be *information* on the proximity of the current solution in relation to the *best* solution (not known).

In many discrete-space problems, there is no *better* or *worse* solution, but the solution is either wrong or right. The objective is rather to find a solution that satisfies a set of different constraints [58, 113, 159, 162]. Understanding the fact that optimising the design of a computer chip can mathematically be exactly the same as optimising gates assignment at airports might, on the other hand, be more challenging. However, the versatility of the fields of application where well-known combinatorial problems are found is also the reason why they are important to study.

Artificial Intelligence (AI) are iterative procedures that combine different operational and organisational strategies based on computerised models in order to obtain high-quality solutions to complex optimisation problems. The AI is concerned with the development of computer programs that emulate the intelligence of humans. AI is concerned with the understanding of human problem-solving strategies, in particular the problem-solving in specific domains by experts.

1.7 BIG DATA

Big data means a large volume of data (structured and unstructured) – that inundates most businesses on a day-to-day basis. It is not the amount of data that is important, and it is what we can do with it that matters. Big data can be contrasted with Small data, thus, it is believed that *"big data is for machines; small data is for people."*

Increasing digitalisation and networking have changed the entire industrial production, management, etc. The volume of data is exploding as the total amount of data in 2005 was 130 exabytes, and that amount had grown to 460 exabytes by 2012. Experts expect the volume of data to grow to 175 zettabytes[7] of data worldwide by

---

[7]One zettabyte is equal to a thousand exabytes, a billion terabytes, or a trillion gigabytes.

2025. In order to analyse and to be able to use such huge volumes of data, one must first develop systems that enable us to understand their contents. One of the preconditions for this is to know how devices and systems function and which kind of sensors and measurement technology can be used to access the most useful data [50].

There are things that are so big that they have implications for everyone, whether you want them or not. Big data is one of those concepts and is completely transforming the way the business will be done in the future and is impacting almost everything! the airport business as well.

As with any step forward in innovation, it can be used for good or abominable purposes. Some people are concerned about privacy, as more and more details of our lives are being recorded and analysed by businesses, and governments every day. Those concerns are real and not to be under-estimated, our belief is that regulations will evolve alongside the technology to protect individuals!

1.8 BAGGAGE FACTORY

From the technology point of view, the automotive and the automatic warehouses industries were and are still ahead in comparison to the airport industry and especially the BHS business. There remains a gap to be reduced! The BHS sector is fairly conservative. Perhaps the risks of operation 24/24 are the basis of this attitude! The investments and the risks involved do not facilitate the task either.

As a matter of fact, most BHS suppliers are also suppliers of the automatic warehouse industry. The similarities between the two industries are obvious. Check-in at an airport is like goods-in at a warehouse; bag storage compares to the storage area in a warehouse, and bag loading has many characteristics of the commissioning area where goods are picked. Learning from each other is necessary. However, there exist some differences between the warehouses and the BHS (Table 1.3).

Table 1.3
Warehouses vs. baggage handling systems

Warehouses	BHS
Pulling System	Pushing System
Order commissioning starts entire process	Check-In starts entire process
All goods go in storage	Some bags are sent to baggage storage area
Random distribution or any systematic	Random distribution or any schematic
Large storage	Small storage
Just in time at system exit	Waiting at gate

The Baggage Factory is an approach that drives up efficiency by simplifying the processes from the moment passengers drop-off their bags at the check-in till their final destination. The Baggage Factory is brought in response to the need for a proven system that can be completed in a shorter time, at lower cost. Like a factory, we need to look at end-to-end solutions. Simply adding technologies to the processes is never the solution. Sure, technologies like RFID, automatic loading, etc. may render BHS more efficient, but we need to look at the processes, their integration and not just the technology. The processes and the work organisation have to be "engineered" as well as the technology, preferably within an integrated approach (see 12.6.1 for the T.O.P. approach).

In general, airports install automated equipment to increase their capacities and improve the passengers comfort. These equipment's use automatic/robotic devices to store bags in storage systems, to track the bags, to load bags into containers, etc. Figure 1.5 shows some of the processes where automation is on-going day-by-day. Table 1.4 gives a short description of the different topics.

In the near future, airports will be able to offer passengers the option of a 100% self-service journey, where human interaction is eliminated thanks to the use of automated, self-service technology at every single step of the journey. Staff will be able to focus on providing excellent customer service without needing to perform basic operational processes.

Automation will allow air travellers to walk freely through the airport processes, from their personal transport to their seats on the plane. We will get to the level where the processes no longer dominate the airport journey and the passengers will have more freedom[8] to do what they want at the airport, whether eating, shopping, working or boarding the plane [171].

Recently, due to the pandemics, the contact-less journey was launched by some airlines. Travellers can book the service up to seven hours before their flight and check-in agents arrive at the location booked at the chosen time. The agent checks travel documents, completes the check-in process and collects the bags. Seat selection can be made at the same time and extra baggage allowance can be purchased if required. Guests who want to board their flights early and receive their bags first can add priority boarding and priority bag tags for an additional fee. Once arrived at their destination, all non-transit passengers who asked for the home check-out service (home bag reclaim) can skip the queues bag-free. Thus, the home check-out service may limit human interaction at the airport as guests are able to head straight to immigration and avoid long lines at the airport, etc. [30].

1.9   HOW THIS BOOK IS ORGANISED

The purpose of this book is not to describe the different airport's concepts, processes in detail, but rather to give an idea of the BHS field and how the different processes

---

[8]It is important to note that the departure time of the flight will always remain a limit to this freedom, as all passengers must complete their boarding procedures before the flight's scheduled departure time.

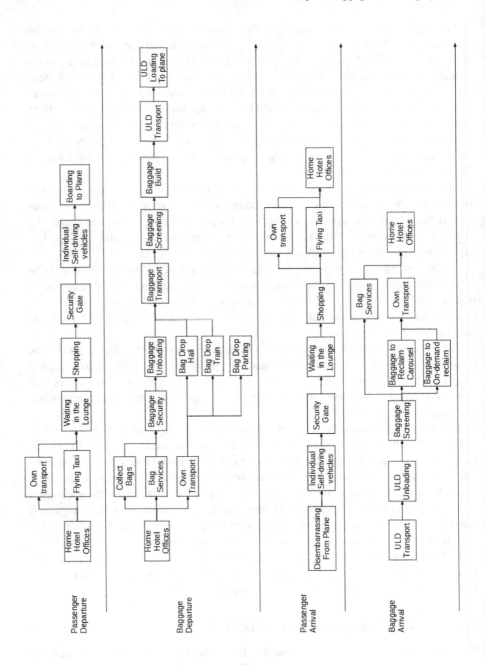

Figure 1.5   Airport Baggage handling – towards full automation

Table 1.4
Processes and automation

Topics	Processes	Description
Passenger Departure	- Home, Hotel, Offices	- collect passengers from home, hotel, etc.
	- Own transport	- passengers use their own transport.
	- Flying Taxi	- future flying taxi between the cities and the airport.
	- Waiting in the Lounges	- frequent flying passengers waiting in lounges.
	- Shopping	- duty free shopping.
	- Security, Gates	- security checks at the gates, before boarding to the plane.
	- Individual Self-driving vehicles	- transport passengers between the lounges and the gates.
	- Boarding to Plane	- finally boarding to the plane.
Baggage Departure	- Home, Hotel, Offices	- the starting points.
	- Collect bags	- collect bags from home, hotels, etc.
	- Bag Services	- different bag services.
	- Own Transport	- passengers use their own means to transport the bags.
	- Baggage Security	- bags security checks.
	- Baggage Unloading	- handling arriving bags to the airport.
	- Bag-Drop-off (airport Hall)	- use the drop-off stations (main hall) to check-in bags.
	- Bag-Drop-off (Train station)	- use the drop-off stations (train station) to check-in bags.
	- Bag-Drop-off (Parking)	- use the drop-off stations (parking) to check-in bags.
	- Baggage Transport	- baggage handling systems (conveyors, AGVs, etc.)
	- Baggage Screening	- bags security checks inside the BHS.
	- Baggage Build	- sort and load bags inside their corresponding ULDs.
	- ULD Transport	- transport the ULDs into the stands.
	- ULD Loading To plane	- load the ULDs into the baggage cargo hold area.
Passenger Arrival	- Disembarking from the plane	- the starting points.
	- Individual Self-driving vehicles	- transport passengers between the planes and the arrivals.
	- Security, Gates	- security checks at the gates, before leaving the airport.
	- Waiting in the Lounge	- frequent flying passengers waiting in lounges (privilege).
	- Shopping	- duty free shopping.
	- Own transport	- passengers use their own transport.
	- Flying Taxi	- future flying taxi between the cities and the airport.
	- Home, Hotel, Offices	- the end points.
Baggage Arrival	- ULD Transport	- unload the ULDs from the baggage cargo hold area.
	- ULD Unloading	- offload bags from ULDs and drop them on conveyors.
	- Baggage Screening	- bags security checks inside the BHS.
	- Baggage to Reclaim Carousel	- transport bags to the reclaim carousel.
	- Baggage to On-demand reclaim	- deliver bags to the passengers on demand.
	- Bag Services	- different bag services.
	- Own Transport	- passengers use their own means to transport the bags.
	- Home, Hotel, Offices	- the end points.

can be improved. By the baggage handling, we mean handling of bags as well as the handling of containers full/empty of bags, etc. Thus, the focus here is mainly on the BHS optimisation and possible process improvements. The aim is to give an idea of what can be achieved with the Baggage Factory approach. This book is divided into three parts as follows:

- In part 1, the main airport BHS' systems are introduced, and the general flows are described as well as the ergonomics, and handling aid solutions. Also, the evolution over the last decades of the baggage handling is described (from conventional to batch building concepts, etc.)
- Part 2 describes the operations of the BHSs. The baggage and ULDs storage and the services that can be delivered to the passengers (customised baggage reclaim, drop-off, etc.) are outlined. Industry 4.0 and its influence on the BHS and its impact on labour, cost, and customer satisfaction, etc. are discussed. The batch build approach based on the pull rather on the push approach is introduced. The automatic packing cell system in which a non-conventional method to load baggage inside containers is presented. The automatic guided vehicles field is one of the emerging ways of transporting bags in airports is introduced. The BHS disaster management questions, and how they should be addressed during the design phase, etc. are also addressed in this second part.
- Part 3 deals with the baggage handling design tools, and the lean manufacturing approach principles as well as the multi criteria decision-aid methods. Then, the optimisation techniques especially the non-deterministic search techniques (AI and optimisation) are described. Next, the simulation which is a powerful tool to help the designers to check new processes and solutions for the BHSs is introduced. We conclude with the future challenges in the BHSs mainly, the IA, the big data and how to use it in an efficient manner.

2 Airports – Baggage Handling Factory

2.1 INTRODUCTION

An airport is composed of landside and airside zones. The airside comprises all areas having direct access to aircraft such as runways, bridges, etc. In front, the landside is composed of the check-in, main-hall, shops, parking, etc. The link between the two zones is the terminal zone which contains the Baggage Handling System (BHS).

The BHS is used to transport automatically the bags through the different processes and sorts them according to their departing flight. Figure 2.1 provides an overview of the main areas of the BHS: the baggage check-in counters, in-feed stations, the baggage screening systems, the baggage sorting machinery, the storage system, the baggage build facilities and baggage claim carousels, etc. [67].

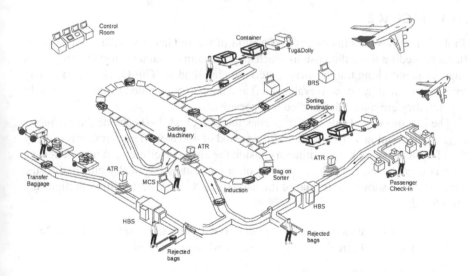

Figure 2.1 A BHS with its components

A non-exhaustive list of equipments used for the BHSs includes Check-in, Conveyors, Diverter (horizontal and vertical), Make-up Carousels, Reclaim Carousels, Chutes, Build Cells, Robot cells, Sorting machinery, DCV, AGV, Totes, ULD Tippers, Screening machines, BSS, ULD Storage, VSU, Lifts, etc.

The baggage handling resources (e.g. the check-in) and the containers handling resources (e.g. baggage tugs) are among the main resources of an airport. The

DOI: 10.1201/9781003432920-2

Table 2.1

In Gauge Bags dimensions (IATA standards)

	Maximum	Minimum
Length	900 mm	200 mm
Height	750 mm	200 mm
Width	750 mm	50 mm
Weight	32 kg	2 kg

management of baggage handling services at airports vary widely across different continents. In many cases, the airport authority provide itself the baggage handling service, while in others, it may be outsourced to independent companies or managed by the airline companies themselves[1]. Ultimately, the ownership and the management of baggage handling services at airports depend on a variety of factors, including local regulations, market conditions, and historical practices.

2.1.1   BAGGAGE

Table 2.1 shows the dimensions and weight of the In Gauge standard IATA[2] bags. Items exceeding these dimensions (such as push chairs, round or unstable bags, wet bags, bags with long trailing straps, etc.) are classified as 'Out Of Gauge' (i.e., bag too long, too heavy, not conveyable, etc.) and have to be handled by other means. In general, they are manually handled by operators.

The License Plate Code (LPC)[3] – alias the IATA code[4] – is a code that defines a baggage item uniquely (worldwide) for a period of 4–6 days. This number is used by the IT systems to look up its itinerary inside the airport facilities[5]. When a baggage item is checked in, the LPC is printed on a tag that is physically attached to the bag. Table 2.2 shows the format of the baggage LPC. Below, the different digits are introduced:

-   A (1 digit) is related to the type of tag : 0 = regular tag, 1 = fallback tag, 2= rush tag, 3 to 9 : free use – meaning depends on the airline.

---

[1]In Asia, baggage handling services are often provided by a combination of airport authorities, airlines, and independent companies. In the Middle East, airport authorities or airline companies often provide baggage handling services. In Africa and South America, the situation may vary depending on the specific country and airport. In Europe, handlers are independent companies, while in US, handling is generally performed by airline companies.

[2]IATA: International Air Transport Association. It publishes standards and recommended practices.

[3]This 10-digit code complies with IATA regulations. Some call it License Plate Number (LPN).

[4]The structure of IATA codes is defined in IATA resolution 740.

[5]Tracking bags across crucial points in their journey inside the airport is one of the goals of IATA resolution 753. This can be also done with "bar-code" bag tags.

Table 2.2
Baggage LPC

1	2	3	4	5	6	7	8	9	10
A	B1	B2	B3	C1	C2	C3	C4	C5	C6

- B^6 contains the company code (a 3 digit number);
- C is the bag code (a 6 digit number).

Figure 2.2 shows the two baggage tag technologies (bar-code and RFID) that are used to track and trace bags inside the BHS. Bar-code tags (Figure 2.2 (a)): are the most popular technology used for the identification of bags. They do require a line of sight. RFID tags (Figure 2.2 (b)) can be read in bulk at distances up to 10 meters. They do not require a line of sight. The tag contains data such as passenger, all the flight information, including passenger destination(s), as well as a bar-code. It can store large amounts of information.

RFID technology offers some advantages over bar-codes, such as the ability to read tags regardless of their orientation, label quality. However, RFID also has some disadvantages, such as the reader's inability to distinguish between tags within its range, making it difficult to ensure that the correct tag is being read. Nevertheless, it can be useful for quickly identifying all bags within a container (provided the antenna is located inside, as the container can act as a Faraday cage).

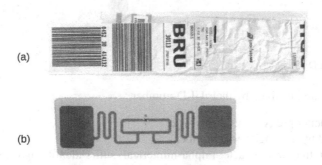

Figure 2.2   Bag bar-code tag (a) and Bag RFID tag (b)

SITA[7] manages an international network that centralises baggage information of the subscribed airports in order to distribute this information to the local automated

---

[6]These three digits are called the "BTIC" (Baggage Tag Issuer Code), as defined in IATA resolution 769.

[7]SITA = "Société Internationale de Télécommunications Aéronautique"

BHS. Baggage information is typically generated by a Departure Control System (DCS) owned by airlines. If the airline has a contract with SITA, their DCS is connected to SITA and sends Baggage Source Message[8] (BSM) to SITA for distribution. Each BSM contains a unique LPC identifier for the corresponding baggage item. To receive the BSMs for checked-in and transfer bags, an airport must sign a contract with SITA. This ensures that the airport can access relevant information for bags processed through its check-in desks and arriving transfer bags.

2.1.2 CONTAINERS

A Unit Load Device (ULD)[9] is a standard container that is used to load bags (see Figure 2.3(a)). It refers to any type of container or pallet used to transport cargo or baggage. A ULD can be of different sizes and is used to ease the loading process into different aircraft types. It allows many items to be bundled into a single unit.

Containers are transported on dollies, and each dolly can transport only one container each time. The ULDs are intended to be loaded into the aircraft and fly along with their loads. Some aircraft do not require bags to be transported in containers and are referred to as non-containerised aircraft, also known as "bulk bags". Bulk baggage is transported on baggage carts (Figure 2.3(b)). A tug (an electric or gasoline-powered vehicle) can pull up to 6 dollies or carts.

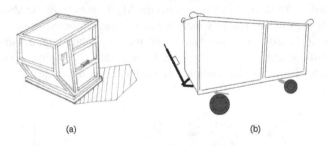

(a)   (b)

Figure 2.3 ULD and baggage cart

The ULDs are identified by their ULD number :

- A three-letter prefix identifies its type,
- Followed by five-digit serial number,
- Ending with a two-character (alpha-numerical) suffix identifying the owner.

For example, AKN 12345 BA means that the ULD is an LD3 with the unique number 12345, and its owner is British Airways (BA). The LD3 is one of the most

---

[8]The BSM contains an identification tag that indicates itinerary of the baggage. This tag provides a unique identification for baggage worldwide as long as the bag has not left the airside of the airport.

[9]ULDs can include both rigid containers like CANs and flexible containers like cargo nets. CAN stands for "Container" and refers to a rigid container used for transporting bags. It is typically rectangular in shape.

Table 2.3
Containers characteristics

ULD Type	Nb Bags	Dimensions
AKE	37	LD3
AKH	30	LD3
DPE	30	LD3
DQF	65	Double length DPE
ALF	90	Double length AKE
...	...	...

used containers and they have the following dimensions ($201 \times 153 \times 163$ cm), with a volume of 4.33 m^3. There exist many variants of the LD3, e.g.:

- AKN: LD3 with forklift holes.
- AKE: LD3 without forklift holes.
- AKH: same base as AKE, extensions on both sides, 114 cm high.

Table 2.3 presents the characteristics of various types of containers.

2.2 BAGGAGE HANDLING

A BHS has at least three main tasks, (1) transport bags from the check-in desks to the departure aircraft parking position[10], (2) transport bags from one aircraft parking position to another during transfers, and transport bags from the arrival parking position to the baggage-reclaim area. Figure 2.4 shows a schematic model of the baggage flows (in and out) inside a typical BHS [55].

On the landside, the bags enter the BHS through the check-in or bag drop-off (A) and leave the BHS through the passenger's baggage claim (D). On the airside, the bags of incoming flights enter the BHS through infeed (C) and the transfer bags are directed to the airside out-flow (E). Bags of an outgoing flight follow the airside out-flow (B). Baggage are sorted and transported between the different areas of the BHS (F). Early bags are stored in Baggage Storage System (G) (see Table 2.4).

Airports have many baggage handling processes check-in, transfer, outbound, inbound, and internal -baggage handling. Domestic airports do not handle in general the transfer bags.

1. Check-in baggage handling: Checked-in baggage items are brought in by passengers arriving at the airports' (landside area). At check-in counters the bags are inducted into the BHS (A).

---

[10] In general, the aircraft's parking is near the gate, but in some airports, there are parking positions that are not linked to a gate and are far from the gates (passengers are transported by bus from/to the terminal).

Table 2.4
Flows between the different areas of an airport

Flow	Description
A	On the landside, bags enter the BHS through the check-in or bag drop-off.
B	Bags of an outgoing flight follow the airside out-flow.
C	On the airside, bags of incoming flights enter the BHS through infeed.
D	Bags leave the BHS through the passenger's baggage claim.
E	Transfer bags are directed to the airside out-flow.
F	Bags are transported between the different areas of the BHS.
G	Early bags are stored in baggage storage.

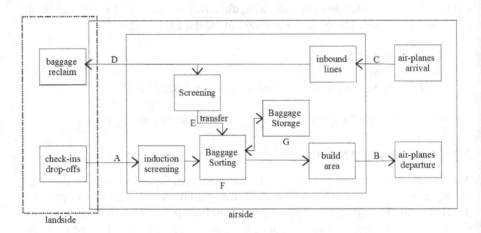

Figure 2.4   Baggage in-flow and out-flow at airports

2. <u>Transfer baggage handling</u>: Transfer baggage are transported to the infeed stations and fed into the BHS (C), from where they are handled as an outbound baggage (E).
3. <u>Outbound baggage handling</u>: Outbound baggage handling include all process steps necessary to input baggage to a departing flight (B). The outbound baggage either comes from passengers through the check-in (A) or transfer bags (E). Bags are sent to a baggage build area where they are loaded into containers and then transported to the outgoing aircraft.
4. <u>Inbound baggage handling</u>: Inbound bags are transported from incoming flights to an in-feed station and transported to the baggage claim carousel where they are picked up by the corresponding passengers (D).

5. <u>Internal flow</u>: Bags are transported between the different areas of the BHS, mainly between the screening area, sorting machinery, manual coding station, baggage storage, customs checks for inbound baggage, etc. (F).

6. <u>Baggage storage</u>: The early outbound bags can be temporarily stored until they become on-time (G).

In order to handle bags at the airport, many resources are needed, first the BHS itself, second the baggage towing vehicles, containers and operators (drivers, handlers, agents, etc.). Chapter 9 presents the resources planning and the needed actors for the ground handling tasks. Figure 2.5 shows the main areas and processes of a typical BHS [55]. The different elements will be described in the following sections.

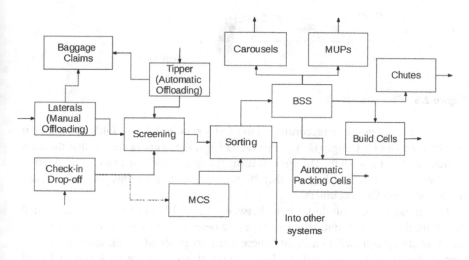

Figure 2.5   BHS – main areas and processes

The BHS influences passengers' perceived level of service at an airport [132]. The size of the BHS and its layout are part of a strategic decision of an airport, and the BHS processes such as the bags routing are operation decisions [41, 91].

2.2.1   IN-FEED LAND-SIDE (A)

In-feed stations (i.e., check-in counters and drop-off stations at the landside) are the access points for bags into the BHS. For each bag, at least one BSM is emitted by the DCS of the airline and is transmitted to the BHS. The BHS uses the BSM information to route the bags to their next handling area, to the storage system, to sorting machinery or to the baggage build area, to the reclaim carousel, etc.

2.2.1.1   Check-In

At the check-in counters, the agent checks if the weight, the dimensions and the convoy-ability of a baggage item are acceptable, then presses the "label/weigh" push-button and the bag is transported to the labelling position, where the agent adds a label to the bag[11]. The length of the bags can be checked using the photocells. In Gauge bags are dispatched and conveyed further on the BHS.

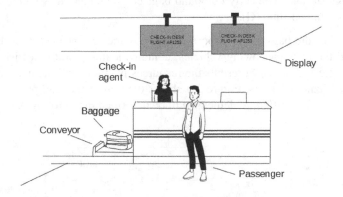

Figure 2.6 Check-in station

If a bag is not in gauge (according to the IATA standards) (see Table 2.1), then it is said to be Out Of Gauge (OOG). For OOG bags, the agent is alerted that the item is prevented by the system. The agent returns the item back to the passenger waiting at the check-in desk. The bag can then be removed from the conveyor and checked in instead at the OOG counters.

In general, a check-in desk opens between 2 and 3 hours before the scheduled departure time of the flight. Often, the service period is fixed and is identical for all check-in desks assigned to a flight. There are also pools of check-in desks shared among flights of the same airline. The opening times may not be identical for all check-in desks assigned to a flight :

- Check-in desks dedicated to Business/Frequent Flyers can have a wider opening time.
- For big flights, depending on the estimated passenger distribution profile, some extra counters are open during the peak of the profile.

2.2.1.2 Drop-Off

More and more, airports are adopting bag drop-off stations in addition to the check-in desks. The normal opening hours of the baggage drop-off are 2–3 hours before the scheduled departure of the flight; the closing times are around 40–60 minutes before scheduled departure. Thus, the passengers can themselves check-in their bags using a drop-off station at an airport.

---

[11] Similar process is carried by the passengers themselves at the "self-service check-in stations" called Drop-off stations.

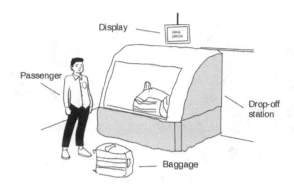

Figure 2.7 Drop-off station

2.2.2 INBOUND FLOW (C)

The unloading process of an incoming aircraft begins once it is parked at the on-block position. Often, bags are sorted within the cargo hold of the plane such that urgent transfer bags with shorter connecting time are unloaded first. The transfer bags are loaded on a vehicle which drives to an infeed station where the bags are fed into the BHS. At the transfer in-feed stations, the bags are unloaded by the operators from the container and put on the conveyor belts. The bags are then transported to their corresponding reclaim carousel. The "non-transfer" bags are unloaded to the inbound line that leads to the reclaim carousel assigned to the flight.

Passengers disembark from the plane and walk to the assigned baggage claim carousel, where they wait to pick up their bags. Once the passengers arrive at the baggage build area and identify their bags on the conveyor belt, they pick them up and finally leave the baggage claim hall when all their bags are claimed [55].

2.2.3 OUTBOUND BAGGAGE (B)

The outbound baggage handling process starts when the check-in agent inducts a bag into the conveyor or when a passenger inducts his bags at the drops-off station. Then, the bags are transported to their build areas. The early bags are stored in the BSS and are released only once the flight's resources have been opened [2].

The bags make their way down via a sorting machinery to their final destination. There, the baggage handlers load the bags onto the containers. Full containers are transported temporarily to the full ULD buffer or straight into the aircraft. Often, the transfer bags (at the origin airport) are loaded into separate containers different from the ones containing the bags that will be heading the destination airport. A bag tag at the sorting station tells the handlers which bags are going where!

For each bag to be loaded, the operator scans the bag's tag[12] and the bar-code of the destination container. If the two scans match, then the bag is loaded into the

---

[12]The scan of the bags is carried only in airports equipped with a Baggage Reconciliation System (since IATA 753); this is mandatory but it is not yet applied in all airports.

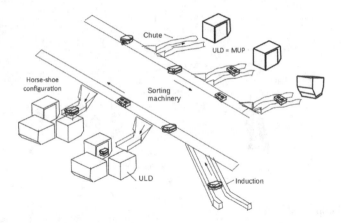

Figure 2.8 MUPs

container, else the bag remains on the conveyor belt of the carousel or is placed on the floor. For security reasons and to reduce the number of mishandled bags, only one flight is assigned to a working station during a period of time [55]. But, in general, for efficiency reasons, more than one flight can be assigned to the same position. Thus, each baggage handler picks up the bags belonging to the flight he is in charge of. So there can be a mix of 4-5 flights on the same working station.

Make up Positions (MUP) are working stations of the handling facility's dedicated to outbound bags. Containers are lined up on parking positions parallel to the conveyor belt (see Figure 2.8). Each working station is equipped with a screen and a scanning device (only in the case of the Baggage Reconciliation System). The screen provides the necessary information about the handled flight such as the number of expected bags, flight's departure time and the destination. The scanning device is used to check whether the bag can be loaded into the container or not.

A carousel is an oval-shaped conveyor belt (Figure 2.9). The carousel capacity is given by the number of bags which can be placed on the conveyor belt. In general, the lane-based handling facilities are often used; each position allows only one flight at a time to be handled. The horse-shoe configuration allows handling several flights simultaneously (Figure 2.8).

2.2.4 BAGGAGE CLAIM (D)

Baggage claim carousels are the final point of a bag's journey from one airport to another. Baggage claim carousels are grouped in the baggage claim halls (situated at the landside), where arriving passengers pick up their baggage (see Figure 2.10). Bags of the same flight are all routed to one of the available reclaim carousels. A connection of an in-feed station to multiple baggage claim carousels offers greater flexibility; a direct connection allows faster processing of the bags [55].

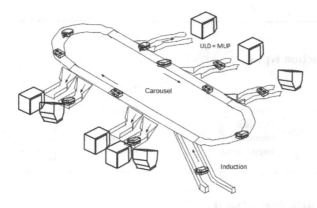

Figure 2.9 Carousel

Airports have more and more knowledge of how reclaim carousel perform their task under congested conditions, especially in hot-seasons. Many software can predict the performance of a reclaim carousel for different demand conditions (i.e., aircraft size, flight, etc.). The characteristics of the reclaim carousel affect the performance of the system [59]. The number of simultaneously served flights is limited by the maximal number of passengers who can stand around the baggage claim carousel.

In general, the passengers disembark from the aircraft and walk into the arrival hall. There, they wait patiently, hoping their bags will arrive soon. Sound familiar? It is of the most frustrating travel experiences for passengers. While other aspects of baggage handling have witnessed substantial improvement in efficiency over recent years, reclaiming baggage has, for the most part, been neglected.

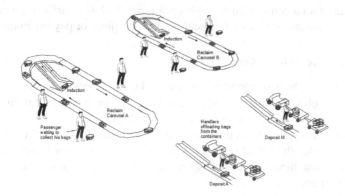

Figure 2.10 Baggage claim carousel

Table 2.5
Flights connection types

From	To
Domestic	Domestic
Domestic	International
International	International
International	Domestic

2.2.5　BAGGAGE TRANSFER (E)

Transfer baggage processes mainly arise at international airports. Most of the passengers coming through it are making a transfer. Therefore, the goal of the HBS is to have the transfer bags keep up with the passengers. Thus, so for the bags to keep up their trip they need to be able to be transported between the aircrafts quickly.

In general, the transferring bags are loaded onto conveyors, where they move through scanning stations and then are routed to their destination. In some cases, transfer baggage are directly brought from an incoming to an outgoing aircraft without traversing the BHS (tail-to-tail principle).

The Minimum Connection Time (MCT) is the minimum time needed to transit at a given airport between two flights. The MCT is based on variables like the airport layout, the security checks, connection type (international or domestic flights), etc. Table 2.5 shows the four possible connection types. When a passenger books a series of one-way flights on his own, there is a risk to arrange a connection that does not meet the MCT requirement. In general, the MCT ranges from 30 minutes to 3 hours, depending on whether it is a domestic or an international flight. Thus, if the passenger misses a flight with a connection that does not meet the MCT, airlines generally do not offer any assistance such as a free seat on the next flight or pay restitution, etc.

2.2.6　BAGGAGE TRANSPORT (F)

Within the BHS, the bags transport can be realised using divers transportation means like conveyors, totes, destination-coded vehicles, etc. These transportation systems have different performances, complexity, flexibility, costs, etc. [87].

At present, conveyor systems exhibit a clear dominance in the BHS market. This is due to the act that the actual selection of conveying technology is determined by criteria such as cost, throughput, sorting criteria, type of conveyor network involved, building constraints, etc.

2.2.6.1　Conveyors

The conveyors of an airport comprise a huge network. The conveyor system can route the bags from the different infeed stations (check-ins, transfer infeed, drop-offs, etc.)

to their corresponding destinations (chutes, carousels, build areas, etc.). Figure 2.11 shows an example of a network of conveyors. There exist two standard configurations for transporting the checked-in bags to the other levels. Bags can either be transported by means of a collector line (see Figure 2.12) or via lifts (see Figure 2.13).

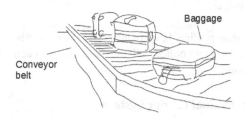

Figure 2.11 Bags on conveyor belts

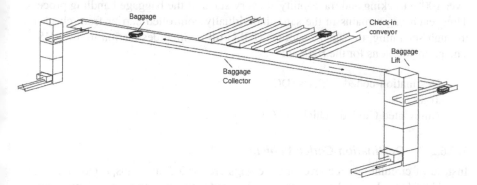

Figure 2.12 Check-in connected to a collector

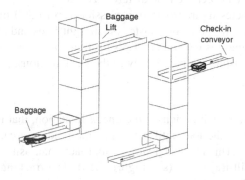

Figure 2.13 Check-in island

Collectors

Collector lines are utilised to transport bags from check-in islands. Each bag is assigned in a designated space (window) on the collector belt. When this window reaches the check-in conveyor, the bag is dispatched. The system aims to ensure that each check-in desk has an equal chance of dispatching bags. Conveyors or lifts are then used at the end of the collector line, to transport the bags to the other levels, where they are screened, etc.

Lifts

Lifts are used for transporting bags from check-in onwards to the other levels. At those levels, collector lines collect all the bags out of each lift belonging to a particular check-in island.

2.2.6.2 Baggage Individual Carrier System

The Baggage Individual Carriers System (BICS) ensures fast and careful handling and delivery of each baggage item to its correct destination. These solutions can deliver 100% tracking and traceability at every stage of the baggage handling process. Thus, each bag remains in the same individually controlled carrier from check-in, through screening, to baggage storage (BBS) till its final destination in the BHS[13]. The main solutions for the BHS are:

- Destination-coded vehicles (DCV)
- Totes
- Automated Guided Vehicles (AGV)

2.2.6.2.1 *Destination-Coded Vehicle*

Instead of circulating on conveyors, the bags are loaded and transported on a high-speed DCV (see Figure 2.14). These self-directed vehicles with on-board intelligence move on static tracks. A control system manages and optimises the different DCVs missions to transport bags between the different zones.

Each DCV completes its transportation mission (up to 500 m) by adapting its speed (up to 10 m/sec) to the layout (faster on straight lines and slower on curves). Each of them communicates with other DCVs upstream and downstream to maintain safe distances and transport the bags safely to their destinations.

2.2.6.2.2 *Totes*

With the Totes technology, it is right at the check-in station, that baggage is loaded into a tote and stays in the tote until it reaches the loading area, even throughout the screening process. This is a much simple approach than using a combination of conventional belts, tilt-trays, etc. (see Figure 2.15). The tracking is also easier as all bags are in RFID-tagged totes. This means routing, diverging, and merging are simplified, and airports achieve higher handling flexibility and safer handling.

---

[13]But up to now, the system that provides the faster throughput is the traditional conveyor lines.

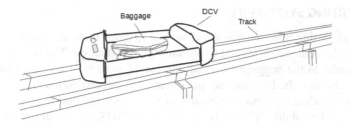

Figure 2.14 Destination-coded vehicle

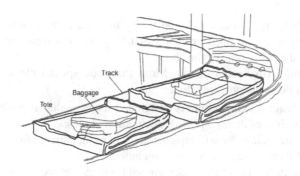

Figure 2.15 Baggage tote system

2.2.6.2.3 *Automated guided vehicle*

Basically, an automated guided vehicle (AGV) is a metal cart (with wheels on the bottom) and a plastic tub on top (see Figure 2.16). The AGV is foreseen to efficiently to transfer bags from one sorting machinery to another. Each individual vehicle in the fleet carries a single bag and determines its optimal route through the airport by itself [183].

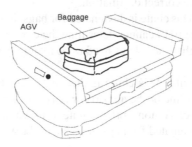

Figure 2.16 Baggage AGV system

2.2.7 SORTING SYSTEMS (F)

The baggage sorting machinery is a mechanical system (in general with a high capacity), which accurately sort baggage to the right destination. The sorting is made possible thanks to the baggage bar-code and the IT systems[14] and the Controls[15]. The IT systems and the PLC instruct per criteria the automated mechanical sorting system as to which is the most suitable destination of a given bag.

A baggage item status, if being processed in the BHS, can be in one of the following situations: early, on-time, hot-rush, late, etc. Its status defines the processes to be carried on it. Thus, there exist many chute types:

- On-time chute (regular case): An on-time chute is a chute in which the appropriate bags are dropped provided that they arrive within the specified opening and closing time of the chute.
- Special chutes (exception cases): More than one special chute could be mapped on the same physical chute:
  + Early chute: An early chute is designated for bags that arrive before the opening time of their respective destination chutes.
  + Late chute: A late chute is designated for bags that arrive after the closing time of their respective destination chutes.
  + Too late chute: This chute is designated for bags that are checked in close to their flight's departure time. These bags may need to be rerouted to a later flight.
  + No BSM chute: This chute is designated for bags that have no BSM assigned yet. These bags cannot be processed by the BHS.
  + No flight chute: This chute is designated for bags that have been checked in for a flight that has been cancelled or has already departed. These bags are then rerouted or returned to the passenger.
  + No allocation chute: This chute is designated for bags that have been checked in but the bags have not yet a destination. These bags are manually sorted to the correct destination.
  + No route chute: This chute is designated for bags that have been assigned to a flight, but the BHS cannot determine the correct route to get the bag to the correct flight. These bags need to be manually sorted to the correct destination.
  + Recalled bags chute: This chute is designated for bags that have been recalled by the airline or the airport for reasons such as security, etc.
  + Non-sorted chute: A non sorted chute (known as "dump chute" or "garbage") is designated for bags that can not be automatically sorted.

There exist many sorting techniques, carousel, tilt-tray sorter, cross-belt, etc. [90].

---

[14]The Sort Allocation Computer is a software which is used to sort bags in BHS.

[15]Controls by misuse of language refer to Programmable Logic Controls (PLC).

2.2.7.1   Carousel Sorter

As part of the baggage handling system, check-in bags are first identified and screened before being conveyed to the sorter. The sorter is a mechanical system consisting of a set of chutes, each equipped with a pusher, which is controlled by a local panel (see Figure 2.17).

To ensure a safe and an efficient handling of bags, the pusher position is monitored by proximity switches, and a photocell is used to verify that the pushing operation has been successful. This means that the area behind the pusher is clear and available for the next baggage item to be processed. The sorter speed can reach 1 m/second, which allows a throughput of up to 2700 bags/hour.

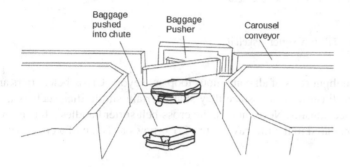

Figure 2.17   Carousel with a baggage pusher – sketch

2.2.7.2   Tilt-Tray Sorter

The Tilt Tray Sorting Machine (TTS) was developed to suit busy airports where fast sortation and handling of different baggage profiles are prerequisites. It consists of a trolley system of linked carriages powered along the sorter track. The bags are transported on trays that are designed to tilt and transfer the bags to different chutes. The trays move along a conveyor system, which enables efficient sorting and routing of the bags. The trolleys support the electrically powered and fully controllable tilting arms, which adjust to handle a wide range of bags, and the bag-carrying trays are robust enough to receive and tip bags of up to 70 kg in heavy working conditions. Figure 2.18 shows a basic sketch of a TTS solution.

2.2.7.3   Cross-Belt Sorter

A cross-belt sorter is a type of sorter that is built on top a of continuous loop conveyor. It consists of a chain of independently operated short conveyor belts mounted transversely along a track. The name "cross-belt" comes from the fact that this belt conveyor runs perpendicular the floor of the loop.

Depending on the number of diverts, and the product being conveyed, it is possible to reach a capacity of approximately 25,000 units per hour. This is amongst the

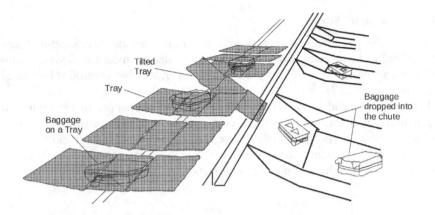

Figure 2.18   Tilt-tray sorter – sketch

fastest throughput rates of all continuous loop conveyors. Cross-belt sorters are often compared to tilt-tray sorters since they share certain similarities such as the frame and drive mechanisms. Nevertheless, the cross-belt sorter handles a bit smoothly the baggage compared to the tilt-tray. Figure 2.19 shows a sketch of a cross-belt Sorter.

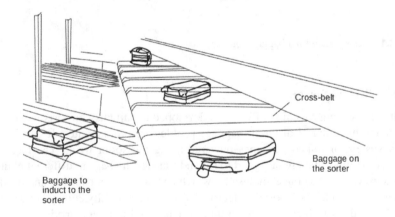

Figure 2.19   Cross-belt sorter – sketch

2.2.8   MANUAL CODING STATION

After check-in, the bags pass through a 360° degrees array of laser scanners (Automatic Tag Reader: ATR) for tag bar-code reading. Most bar-codes are read automatically provided the tag is printed correctly. A few per cent of no-read tags will need manual identification by an operator at a Manual Coding Station (MCS)[16].

During its journey in the BHS, if a bag's label is unreadable, the bar-code scanner returns a "no read"[17], and, if the Video Coding approach (see below) is also not successful, the bag is directed to the MCS, which is in general located downstream of the ATRs. If successful, that bag is then back inducted into the system. Otherwise, the bag is sent to the dump chute for a deep search.

Figure 2.20 depicts the process to identify the bags at the MCS, and can be summarised as follows:

- First, the operator tries to scan the bag tag with a hand scanner, which allows a better reading than with an ATR (operator can flatten the tag, retry several times from different angles, etc.)
- If the bar-code is unreadable, then the operator can enter the 10 digits of the LPC via a keyboard or via the touch screen of the MCS.
- And only if this LPC is not known by the SAC (which means we have a NO BSM error case), then the operator can manually select the flight (by reading what's written on the tag) and assign the bag to this flight. Thus, the SAC (or the MCS) creates a "proxy BSM", which only contains the minimum to sort the bag (=flight number, flight date, flight destination and LPC).

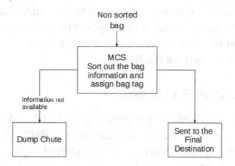

Figure 2.20 Manual coding station process

The MCS was developed to maximise the flow of bags to be encoded. A capacity of 15 bags per minute is frequently reached. These results are obtained thanks to the rational ergonomics of the station and the minimum of information the user has to encode, which drastically reduces encoding errors.

---

[16]Some airports and Baggage habdling system's suppliers use Manual Encoding Station (MES) rather than MCS.

[17]The following bags can be identified at the MCS : no-read bags, out-of-tracking bags, no BSM information, no flight information, etc.

With respect to the hub airports, one of the biggest hurdles is the processing of transfer baggage, especially bags with damaged or unreadable bag tags (between 3-10% can be unreadable). Conventionally, the no-read tag bags are diverted to an MCS. Thus, some transfer bags may be sorted twice, introducing delays as well as having a direct impact on the sorter capacity. To address this issue, one of the solutions is to use the Video-Coding System [68].

2.2.9 VIDEO CODING SYSTEM

The Video Coding System (VCS) system relies on vision systems installed in the BHS which send bag images (on the fly) to an operator in the control room, to zoom in to view no-read tags, and to encode the missing information of any unreadable bag tag. This allows the baggage to remain in motion within the BHS and eliminates the need for the handlers to physically manipulate the baggage. It means that bags will no longer need to be redirected to manual station, minimising both the physical work for operators and bottlenecks during peak times. If the operator is unable to successfully identify a bag using the VCS (within a certain time), this bag is then sent to the MCS for manual identification [92].

Another solution is to use cameras (e.g., photo tracking) either directly at the system input locations (check-in desk and transfer input line) or in combination with the automatic tag readers. The bag images enrich the management information. Therefore, if for a reason, a baggage is lost in the system, the bag information (tracking information, etc.) combined with the images can be used to find lost bags. Capturing images of bags is particularly useful in cases where bags are damaged. By reviewing the images captured at the time of check-in, airport staff can determine if a bag was already damaged prior to entering the BHS.

2.2.10 HOLD BAGGAGE SCREENING

Security at airports has received more attention in the last decades. A checked-in bag is first screened before it is forwarded further into the BHS. Depending on the security standards of the originating airport, the transferred bags may be also screened before they are forwarded through the BHS. Net, standard 2 and standard 3 are presented.

2.2.10.1 Standard 2

Once the bag tag is identified by the ATR, the baggage is sent to the screening area. First, the bag is automatically analysed by the inspection machine (level 1). In case of doubt, the image is pushed to the screen of an operator (level 2).

The "level 2" screening process is conducted by operators viewing the X-ray images at one of the "level 2" screening stations, that are located in a separate HBS screening room. The security controller has 30–60 seconds to re-qualify the baggage according to his analysis of the images. All bags which pass the screening process are sorted through their next destination (chutes, divert, etc.). All the bags that fail

Table 2.6
The standard five-level screening steps (Standard 2)

Level	Description
Level 1	Automated evaluation of the X-ray image by the X-ray machine.
Level 2	Operator analysis of the level 1 image at remote stations, carried out whilst the bag continues in transit.
Level 3	An in-depth analysis of the original level 1 images at a separate station or screen the baggage using a different X-ray technology.
Level 4	Re-unite the passenger and its bag and carry out a manual search.
Level 5	If the passenger cannot be found, then the bag is considered a threat and dealt with accordingly.

"levels 1 and 2" screening are dispatched to the "level 3" Computed Tomography (CT) screening process.

Level 3 receives all bags for which a "level 2" decision is not reached in time, and all those that are lost by the tracking system during this part of the transfer process. Bags which subsequently clear level 3 security are re-introduced into the sortation system, whilst any that fail level 3 security are diverted from the system to a remote pick-up point for passenger reconciliation (level 4) and further inspection.

At level 4, the bag is inspected by an operator who carefully re-examines the images taken by the inspection machines. The controller has an additional 60–75 seconds to carry out further analysis and possibly re-qualify it.

It is only when the controller has doubts that the baggage is sent to a reconciliation room for a manual search and the passenger is called and is requested to open his bag. Table 2.6 summarises the standard five-level screening steps (Standard 2).

2.2.10.2   Standard 3

The technology used for Standard 2 has mainly been X-ray machines. With Standard 3, most of the machines will be CT, which create a 3D image of the baggage that allows for screening explosives easily. The modern CT machines have he advantage of clearing about 80% of all bags versus just 70% for the traditional X-ray. Table 2.7 summarises the standard three levels of screening steps (Standard 3).

2.2.11   BAGGAGE STORAGE SYSTEM (G)

While the local bags enter the BHS through check-in counters, the transferred bags are brought from arriving plane to infeed-stations, where the bags are transferred into the BHS. Each outgoing flight has its own time-slot, which begins 1–3 hours before its scheduled departure time (STD) and ends 15–20 minutes before the STD.

Table 2.7

The standard three-level screening steps (Standard 3)

Level	Description
Level 1	Automated evaluation of the 3D image by the CT machine.
Level 2	Alarm 3D images sent to the Operator for analysis.
Level 3	Final examination, If the passenger cannot be found, then the bag is considered a threat and dealt with accordingly.

A bag for an outgoing flight that arrives at the BHS during the flight's baggage handling period is said to be on-time. On-time bags are in general directed to their assigned handling facility, where they are sorted and loaded into containers.

Baggage that arrives before their flight's time-slot are said early bags and are directed to the storage system, where they are stored until their time slot is reached. Each stored bag is later-on released from the storage system and sent back to the sorting system (see Chapter 5). Figure 2.21 shows an example of baggage storage system.

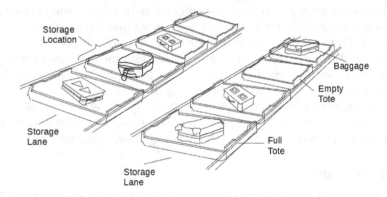

Figure 2.21 Baggage storage system

2.3 CONTAINERS HANDLING

In general, bags are transported via carts and manually loaded into the aircraft hold at the parking position. Nevertheless, many big airports use containers (ULDs and CANs) that are transported on dollies[18]. Each dolly can transport only one container

---

[18]Dollies are standard-sized flat-bed trolleys or platforms, with roller bars and/or ball bearings for easy loading and unloading of ULDs.

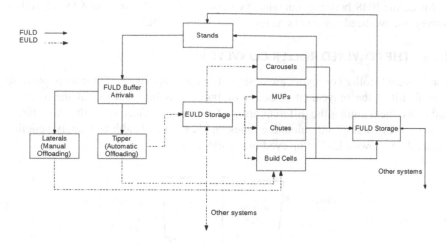

Figure 2.22 Empty and full ULDs flows

each time. A Tug, which is an electric or gasoline-powered vehicle, can tow up to 6 dollies or baggage carts. Figure 2.22 shows the main processes of Empty (EULD) and Full ULDs (FULD), while Table 2.8 gives a short description of the different areas and processes.

Table 2.8
BHS areas and ULDs processes

Area	Description
FULD Buffer Arrivals	Full ULD are buffered before being transported to the baggage hall.
Laterals (manual Offload)	Conveyor lanes where bags are offloaded from the ULDs.
Tipper (automatic Offload)	Automatic offloading of bags from the ULDs (see 3.4.5).
Stands	Stands where the aircrafts are loaded and offloaded.
EULD Storage	Empty ULDs are stored before being transported to build areas.
Carousels	Handlers pull bags from the carousels and load them into the ULDs.
MUPs	Handlers pull bags from the MUPs and load them into the ULDs.
Chutes	Handlers pull bags from the chutes and load them into the ULDs.
Build Cells	Load (manual, semi-automatic, or automatic) bags into the ULDs.
FULD Buffer Departures	ULDs are buffered before being transported to the stands.

Inside the BHS building, containers can be transported using Tug&Dolly, roller-conveyors, overhead conveyors, autonomous vehicles, etc.

2.3.1   THE POWERED ROLLER CONVEYORS

The Powered Roller conveyors are designed to transport ULDs. Their main purpose is to facilitate the movement of heavy or bulk loads into and out of the aircrafts and move them among the different areas of the airport. Thus, they offer significant reductions in loading and unloading times and have huge health and safety benefits. Figure 2.23 shows a ULD conveyed on a roller conveyor.

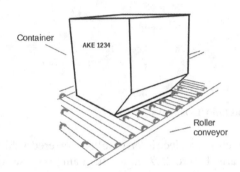

Figure 2.23   Roller conveyors

2.3.2   THE CARGO HANDLING AND STORAGE AREA

The ULDs are not suited to any kind of stacking, nesting solutions and have many different sizes and shapes. More, due to the huge pressure on space at and around most airports, arranging a ULD storage is a challenge. Thus, as part of the overall airport operations, a local cargo area can be used to handle containers. These cargo areas can also be used to store the ULDs safely. Figure 2.24 shows a sketch of cargo storage area.

2.3.3   THE ULD TRANSPORT

The autonomous vehicles transport full/empty ULD from/to the baggage hall from the build area to aircrafts parking positions. Early full ULDs as well as empty ULDs, can be stored inside or outside the main baggage hall.

Figure 2.25 depicts two solutions can be used for the transportation of full and empty ULDs within an airport: (1) autonomous vehicles and (2) overhead conveyors. The solutions are a part of the airport's advanced baggage handling system, designed to enhance the efficiency, the accuracy and the safety of the handlers in the transportation of ULDs. The use of vehicles helps to reduce the need for manual labour and increases the speed of baggage transportation. More details about this subject can be found in Chapter 6.

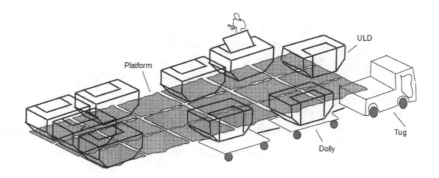

Figure 2.24   Cargo storage area

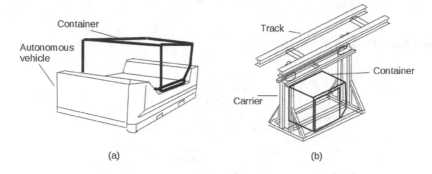

Figure 2.25   Autonomous Vehicle and Overhead conveyor – examples

2.4   IT AND CONTROLS SYSTEMS

The expected functions of a BHS cover: the check-in of baggage; sorting and trans-fer; pick-up of baggage by passengers; buffering of baggage in a holding area; facil-ities for arrival baggage; and the provision of special workplaces, etc. The physical configuration of the control system is often found in the form of a two-level concept.

The first level (Controls) includes the field elements such as sensors and actuators, while high-performance group control units coordinate the switching, signal report-ing, data acquisition, and monitoring of the kinematics of the conveyor elements.

The second level (IT) consists of supervisory and material flow computers re-sponsible for higher-ranking operations entailing open-loop control, management, operator control, and visualisation of the sections of one or several linked baggage conveying systems (see Figure 2.26).

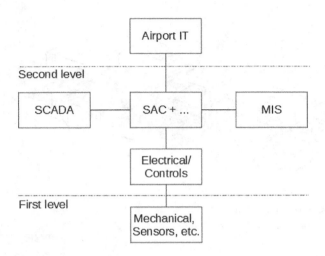

Figure 2.26   IT, Controls, and Mechanical

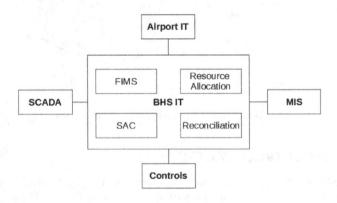

Figure 2.27   IT systems

2.4.1 IT SYSTEMS

Figure 2.27 shows a high level architecture of an IT system. It is composed of many elements such as the Sort Allocation Computer (SAC), the Resource Allocation, the Flight Information Management System (FIMS), the Baggage Reconciliation System (BRS), Management Information System (MIS), etc.

2.4.1.1 The Sort Allocation Computer

The tracking of bags does not begin at the check-in desk, but rather, it begins after the first ATR[19]. The ATRs are connected to the PLC which will pass the bar-code read by the ATR to the IT system (Sort Allocation Computer: SAC) and require a response in the form of sorter number and chute number that the bag should be routed to.

The sort algorithm is based on the matching of the received data. The sort is done for a requested bar-code from which the SAC tries to find the matching received BSM[20]. The flight information included in the BSM is used by the SAC retrieves the allocation information for the bag at hand using the different matching criteria (date, flight, class, destination, handler, etc.). Unread bags are routed to an MCS to be manually scanned and identified by operators (see 2.2.8).

A sort destination is a logical identifier used by the Flight Information System (FIS) and the SAC to address a position on the sortation system. This position can be a final exit (chute or lateral) or it can be the next position on the sortation system where a sort decision has to be taken.

To perform the sortation, the SAC exchanges different types of data :

- BSM data received from the DCS's and stored as a bag information.
- Flight and Allocation information from the FIMS.
- PLC sort requests from the PLC.
- Status information to the MIS.
- Sort replies to the PLCs.
- BPMs messages for the BRS.
- etc.

In general, the SAC supports a multi-criteria sortation (in the order of importance): date, LPC, airline, flight, class, etc. These criteria enable the SAC to determine the most suitable chute:

- Chute(s) allocated to the flight (for secure and non-secure bags).
- Chute(s) for early bags per handler, per sorting machine, and per airline.
- Chute(s) for late bags per handler, per sorting machine, and per airline.
- Chute(s) for unknown bags.
- Chute(s) for internal transfer.
- etc.

The SAC allows also multi-allocation: (1) One chute for several sortation criteria, (2) Several chutes for the same sorting criteria.

---

[19]The ATR is an array of bar-code scanners arranged 90° or 360° around the conveyors. This device can scan the bar-codes on about 99% of the bags that pass by.

[20]The BSM (Baggage Source Message) contains the flight details and passenger information. It is generated and sent by the departure control system, to the BHS IT system. Each baggage item has its own BSM valid for 24 hours, till it leaves its destination airport. The automated baggage handling system scans the bar-code of the bag and sorts it accordingly.

The IT system can trace all the operations carried on each baggage:

-   Reception of BSM,
-   Read LPC code by the ATR,
-   Sorted by the SAC,
-   Sorted at the MCS,
-   Dropped into a storage system,
-   Dropped in chute,
-   etc.

This information is collected in a central database and can be set available for all tasks on request.

2.4.1.2   Resource Planning

Based on rules and constraints, the resources are allocated (planning) to inbound and outbound flights based on a strategy defined by the supervisor. The planning is then transmitted to the SAC system. The resource planning is realised in the following conditions:

-   Anticipating from the generic flight data (exp. per season) and from the data provided by the Flight Information Distribution System (FIDS).
-   In real-time with updates from FIDS by considering the modification concerning the flights already allocated and the new flights to be allocated.

For more details about the resource allocation principles, refer to Chapter 9.

2.4.1.3   Baggage Reconciliation System

The Baggage Reconciliation System (BRS) supports the reconciliation of passengers are used to ensure a full tracing of the bag from the moment the passenger has checked in, until the bag is put on the reclaim belt in the arrival airport, as requested by IATA resolution 753. The BRS receives information from DCS and FIMS systems. In addition, the system receives operational input from the baggage loading process.

The BRS are in fact Baggage Tracking and Reconciliation System (BTRS) that also receive other kinds of IATA 1745 messages, mainly Baggage Processed Message (BPM) but also Baggage Unload Message (BUM) and produce other IATA message like Baggage Not Seen Message (BNS), Baggage Manifest Message (BMM), etc.

The BRS supports the following functional aspects:

-   Process flight and allocation information received from the FIMS.
-   Process BSM information received from the DCS.
-   The handler associates stowage devices to outbound flights.
-   The handler associates baggage items to containers.
-   The creation of flight manifest reports.

- The creation of ULD manifest reports.
- The handling of baggage unload instructions.

According to the baggage reconciliation process, a flight can be:

- Unsafe (there is at least one reconciliation issue to consider).
- Boarding (boarding in progress).
- Locked (boarding finished).
- Away (flight take-off took place).

2.4.2 PROGRAMMABLE LOGIC CONTROLLER

The PLC steers the mechanical infrastructure of the BHS, including the ATR, baggage screening machines, conveyors, etc. When the LPC of a baggage item has been scanned by an ATR, the PLC sends a sort request to the SAC to obtain the baggage next destination. The PLC's determine the sort trajectory to be followed. A sort request is usually completed by a "drop": the PLC informs the SAC in which chute the bag has been dropped.

In modern BHS installations, the PLC plays an important role in tracking and monitoring the movement of bags. The PLC sends "bag seen" or "bag position" messages when the bag passes critical points of the sorting installation. Additionally, the PLC transmits the inspection result to the SAC after the bag passes through an inspection machine.

2.4.3 SUPERVISORY CONTROL AND DATA ACQUISITION

Supervisory Control and Data Acquisition (SCADA) systems are commonly used in airports for condition and status monitoring of BHSs [9]. SCADA enables problems to be shown via a visualisation system, making it part of the daily routine to manage failures, and minor problems are cleared in minutes.

BHS are continually monitored by the SCADA operators in the control rooms, from where they communicate failures to the field maintenance technicians. The user interface provides the operator with system-related information, e.g. bag information, security status, location, number of carriers per zone, number of ULDs waiting, number of bags loaded, etc.

The SCADA system is a convenient way to start and stop the BHS installation with one click at the beginning and end of the day, respectively. Without the SCADA, the technicians would have to go to each control cabinet in the BHS and press a button to start each section of the installation.

Failure in a BHS can be divided into "avoidable" and "non-avoidable". Non-avoidable failures are those that occur on a daily basis, with the most common being bag jams when baggage is caught on side guards or belts. Indeed, given the nature of the systems (curves, inclines, declines, chutes, etc.) and the huge variation among bags, such failures are common, and airports simply live with these failures and have staff available to clear the problems as quickly as possible. It is advised to solve "avoidable failures" by following the preventive maintenance principles and propose solutions to problems [95].

2.5 BAGGAGE FACTORY

In year 2000, was introduced the first Baggage Factory (BF) to the airport business under its new generic "The Bag Factory"[21] principle. It was integrated into an imposing building, this centralised facility is the first of its kind dedicated to the comprehensive processing of transfer baggage to be installed at the airport. It was at that time able to handle over 5500 bags per hour, quite a challenge!

A key module (at that time) within "The Bag Factory" solution was the Bag2000 system (IT and Controls systems). This essentially encapsulates all the control, monitoring and information systems required for baggage handling and includes a SAC and an MIS. Another module was 100% Hold Baggage Screening (HBS), which was developed and packaged with smart screening machines [122]. Other easily integrated system elements include bag separation, bag suitability check, auto tag reading (ATR), manual coding, bag tracking, etc.

Nowadays, the BF includes more concepts and tools than before. Indeed, the baggage handling process can be streamlined by using advanced handling techniques from the automotive industry and the automated warehouses to provide the flexibility to meet the high peak demands experienced by major transfer hubs. Therefore, for big airports, new tools and concepts have been integrated to the BHS such as :

-   Multi-purpose baggage storage
-   Batch build process
-   Autonomous Vehicles
-   Reclaim on Demand
-   etc.

2.5.1 MULTI-PURPOSE BAGGAGE STORAGE

In general, big airports need many chutes to handle the baggage build process. At the same time, it is more and more difficult for an airport to find the budget and space for new chutes in an already cramped environment? The purpose of the conventional baggage storage systems is to have a place in the airport to store stopover or transfer bags and early checked-in bags. Baggage storage systems are assets that airports can not use them only as early baggage storage, but paired with the right storage and baggage handling system technologies, and they can be transformed into multi-bag storage spaces.

Thinking of baggage handling in terms of factory or warehouse systems can be useful, as warehouse looks very well to airports in respect of the way baggage can

---

[21]The Baggage Factory was first developed by Logan Fabricom Airports Group at the Gatwick's 'Transfer Baggage Facility' TBF to manage the reception, identification, coding, security screening and automatic sorting of all of BA's transfer baggage arriving at the North Terminal of the airport [122].

ideally be stored and handled. An airport can store baggage individually, and in the same way as in warehouses, then pick and sort baggage in batches when there are enough bags to make a batch. Using the right batch build algorithms to control the flow of bags between the storage and build areas, airports can become effective "bag factories".

Thus, this storage and batch building is one of the solutions for airports experiencing crammed baggage halls, with no space to expand the make-up area. With batch building, baggage storage systems can be transformed into multi-bag stores, rather than being used just for storing early bags [130].

2.5.2 BATCH BUILDING

The principle of batch building comes from the warehouse practices, where goods can be stored and retrieved in batches. Batch building is a concept used to optimise the loading of bags for departing flights. Once a bag has been checked and screened, it is then stored in the BSS. Once a certain amount of bags has been reached, the control system then alerts the operator that a batch is ready for loading. An operator then "pulls" the batch of bags destined for a given flight from the storage (rather than the random pushing of baggage through a conventional system). The operator can then release the batch and load it effectively within minutes into the container (see Chapter 8) [14, 129].

Batch building means that the chutes must no longer be open for 2-3 hours waiting for bags. It also means that an operator no longer needs to walk back and forth to move any occasional bags over to the container. The operator simply handles all the bags for one departure at defined period of time (in less than 10 minutes per container).

2.5.3 USE THE HEIGHT OF THE BUILDING

Historically, bags were typically stored on the floor, in storage lanes or loops. Today, more and more the racking and the crane systems are used to store bags in many airports. These systems are similar to the conventional high-bay racking systems, where bags are stored individually on shelves.

The advantage of the racking is that for the same storage capacity as the conventional storage lane, it occupies substantially less space. As an example, conventional storage lane system can hold 800 bags using 1200 m^2, while a racking system for a capacity of 800 bags but occupies just 730 m^2, saving around 40% of the valuable floor space [130].

2.5.4 RECLAIM ON DEMAND

Travellers want to spend less time waiting and have greater visibility into the next step of their journey and, especially, the arrival of their bags. Nowadays, some airlines use bag-tracking applications to notify passengers of an estimated collection time so that they do not have to wait unnecessarily while watching everyone else's bag turning around the conveyor belt [69].

On-demand baggage storage is possible, along with the storage for reclaim on-demand bags that need to wait in the system until the passenger calls for it. The Reclaim on-demand is an automated system designed to make easy the baggage localisation for arriving passengers through a messaging application.

2.5.5   BAGGAGE INDIVIDUAL CARRIER SYSTEM

The Baggage Individual Carrier System (BICS) technology is based on carriers used to transport baggage and can be tote-based or cart-based (see Chapter 6). In the BICS, each bag is loaded individually into a carrier and stays within the carrier from check-in desks till its final destination.

The BICS technology will become more and more a standard for big airports because these individual carriers can ensure complete track and trace of bags. The BICS technology is in general faster than conventional conveying systems, and it is able to cope with swift transfer connection times, energy efficient and experiences fewer system jams. As BICS-based systems are fully automated, they can be interfaced with "Reclaim on Demand" solutions [14, 130].

2.6   CONCLUSIONS

In the near future, the Baggage Factory will have more and more features such Artificial Intelligence blocks, Big Data, Vision systems robots, etc. The Baggage Factory is the future of BHS of big airports, and it is characterised by new levels of controlling, organising, and transforming the entire value chain of services, resulting in higher efficiency and flexibility [14].

3 Ergonomics

3.1 INTRODUCTION

Airport baggage handling, as in many other industries such as intra-logistics, is a high-risk job for musculoskeletal disorders caused by handling heavy weights and high frequency. Thus, more than 100,000 pieces of bags are processed manually each day by baggage handlers at many international airports, in which some of the bags weighing more than 30 kg.

It is known that having to frequently handle heavy items creates high levels of physical stress for baggage handlers. Some factors such as size, grip and the baggage material, etc. can also have a negative impact. Since the handlers must repeatedly reposition the bags at awkward angles and to strict deadlines, they are more likely to suffer work-related injuries.

When considering baggage throughput rates, it is known that manual handling is certainly quicker than using any handling aid device during the peak periods. Nevertheless, this rhythm is not sustainable throughout the whole day, month, year, etc. The fatigue and injuries are inevitable in such environments. The baggage processing area has the highest percentage of injuries and effects. Ergonomic studies have been conducted in airports since the early 1990s. Statistics prove that it is getting worse as the number of passengers (bags) increases. Several organisations provide recommendations and best practices to the baggage handling industry in workplace safety and health [51, 173].

The Health and Safety Executive (HSE) [82] guidelines are based on a work frequency of 30 operations per hour (one lift every two minutes), and for faster work requirements, abatement factors are applied. Typically, the transfer of bags from a chute to a container is accomplished through manual handling, occurring at a rate of 4 bags per minute on average.

3.2 THE ERGONOMICS ANALYSIS

3.2.1 THE BIOMECHANICAL ANALYSIS

The biomechanical analysis has contributed to enhance the knowledge of the underlying causes of movements. The biomechanics have enabled a precise quantification of motor strategies in order to optimise, maintain or develop high-level human performances while preventing musculoskeletal disorders in ergonomics [98, 135].

The development of sensors and recording technologies has contributed to the utilisation of biomechanical assessments in many fields. In general, the measurements during a biomechanical analysis provide temporal, load and position information. These measurements can then be examined in relation to muscle fatigue development as it is suggested to be a precursor of injuries. The kinematic recordings are

DOI: 10.1201/9781003432920-3

used as an input or as a validation tool to the applied models investigating the human movements [135].

In general, the measurements of the equipment are taken, and videos of its simulated operation are made for later analysis using the Quick Exposure Check or Manual handling Assessment Charts tools. The utilisation of these tools, can enable a comparison to be made of whether an equipment is likely to be beneficial (or not!) in terms of reducing musculoskeletal risks to the operator.

3.2.2 THE ASSESSMENT TOOL ANALYSIS

3.2.2.1 The Manual Handling Assessment Charts

The Manual handling Assessment Charts (MAC) tool was developed to help identifying high-risk workplace manual handling activities. It is used to assess the risks posed by lifting, carrying, etc. and to understand, to interpret and to categorise the level of risk of the various known risk factors associated with the manual handling activities. It incorporates a numerical and a colour-coding score system to highlight high-risk manual handling tasks [77, 138].

Table 3.1 presents an example of a MAC risk analysis (subjective study) carried out for a task under the two conditions with and without the use of the mechanical aid. This risk assessment identifies certain aspects to be of concern in the activity. The concern is with the weight of the bags when lifted, the frequency at which they are handled, and the poor postures of the operator during handling, etc.

These issues are significantly reduced (if not completely eliminated) if the operators can use a mechanical aid device. The MAC tool scores suggests that the introduction of the mechanical aid device would result in a significant reduction

Table 3.1
Results of the MAC analysis – example

Risk factors	Without aid	With aid
Load weight and lift/carry frequency	5	0
Hand distance from the lower back	3	0
Vertical lift region	3	0
Trunk twisting/sideways bending	2	1
Postural constraints	1	1
Grip on the load	2	0
Floor surface	1	0
Other environmental factors	1	1
Total score	**18**	**3**

Table 3.2
Results of the QEC analysis – example

Risk factors	Without aid		With aid	
	Task Score	%	Task Score	%
Back: posture and weight	10		4	
Back: posture and duration	8		4	
Back: duration and weight	10		6	
Back: frequency and weight	8		4	
Back: frequency and duration	6		4	
Back – Total Score	42/56	75	22/56	39
Shoulder/arm: height and weight	8		6	
Shoulder/arm: height and duration	8		6	
Shoulder/arm: duration and weight	12		6	
Shoulder/arm: frequency and weight	8		4	
Shoulder/arm: frequency and duration	6		4	
Shoulder/Arm – Total Score	42/56	75	26/56	46
Wrist/hand: repeated motion and force	6		4	
Wrist/hand: repeated motion and duration	6		4	
Wrist/hand: duration and force	10		6	
Wrist/hand: wrist posture and force	6		4	
Wrist/hand: wrist posture and duration	6		4	
Wrist/Hand – Total Score	34/46	74	22/46	48
Neck: neck posture and duration	6		6	
Neck: visual demand and duration	6		4	
Neck – Total Score	12/18	67	10/18	56
Work pace	1/9	11	1/9	11
Stress	1/16	6	1/16	6
Total score	132/ 201	66	82/201	41

to the manual handling risk exposure with respect to back-injury risk (see JPR Ergonomics[1]).

3.2.2.2 The Quick Exposure Check

The Quick Exposure Check (QEC) is an ergonomic risk assessment tool for measuring the ergonomic risks to the upper and lower parts of the musculoskeletal system. Bio-mechanical assessments can be done for all the regions of the musculoskeletal system, especially shoulder moments and moments about the low back [38].

Table 3.2 gives an example of the QEC risk analysis for the *manual handling* of bags without the mechanical aid, together with the scores for the task when

---

[1] https://www.jrp-ergonomics.co.uk/

undertaken using the handling aid device. The QEC scores suggest a notable reduction in risks when using the mechanical aid devices when compared with the manual activity (except the back).

The scores represent a hypothetical relationship between the increased level of exposure and a potential health outcomes. The aim of an intervention is to reduce exposure scores, and the higher the score, the higher the risk related to the task.

3.3 MECHANICAL AIDS FOR BAGGAGE HANDLING

The ideal loading aid devices should be designed from an ergonomic perspective. To achieve this, a human-centred approach is necessary, with the aim of developing a solution accommodating the baggage handler's needs. They must be able to work intuitively and easily without lifting heavy loads or having to move bags far away.

Loading aid devices should deliver reliable throughput rates and a consistent performance. Several variants are available on the market, and were primarily designed to reduce the ergonomic risks. Ideally, an ergonomic loading aid device should also increase the throughput (number of bags handler per hour!).

As part of the continuous improvements of baggage handling facilities at airports, many mechanical handling aid devices (e.g., loaders, unloaders, lifters, etc.) are being and have being trialled (and many of them are nowadays used daily) in airports. These devices have undergone several prototype developments following engineering and ergonomics reviews [31, 137, 147, 182]. Many tools and technologies have their origin from the automotive and warehouses businesses.

3.3.1 THE HOOK

The hook is one of the simplest tools to help operators to handle heavy and/or hard-to-reach bags (see Figure 3.1). The hook consists of a long handle with a curved or angled end that can be used to lift or pull bags without the need for direct physical contact. Thus, the operators can use a hook to bring the bags close to them. The hook is a useful tool for maintaining operator safety and reducing the risk of injury when handling heavy or awkwardly shaped bags.

3.3.2 VACUUM TECHNOLOGY

Figure 3.2 shows a vacuum vertical loading aid unit being used to hold bags securely in place during transport, reducing the risk of damage. This unit is an important tool for ensuring the safe and efficient loading and unloading of baggage. This loading aids are used to grab bags from above by a vacuum. They are suitable for containers with open tops (like Wessex or open-top ULDs)[2]. Vertical loading aids are popular as they require a low investment and can be retrofitted, but they are also cumbersome [110].

---

[2]Unfortunately, not all the ULDs are open from the top.

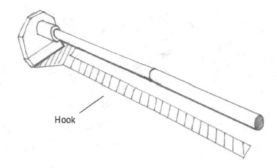

Figure 3.1   Baggage hook

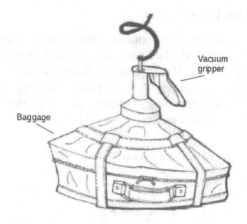

Figure 3.2   Vacuum technology

3.3.3   SCANNER GLOVES

One of the main challenges in a manual scanning operation is that the handlers are not permitted to lift a baggage with only one hand. Thus, the handler has to pull the bag, use the scanner to read the tag, puts away the scanner, and then position correctly the bag inside the container. This process is repeated for each bag to be handled, this creates a laborious movements and processes for the handler.

Figure 3.3 shows a sketch of a scanner glove that can replace the use of conventional bar-code technology. The scanner is integrated into a glove, and the operator triggers the scanning function by pressing (per example) the thumb and first finger together. The utilisation of the glove, leaves both hands free for work and saves additional movements – to pick up and put down a scanner. This makes the working routines more ergonomic.

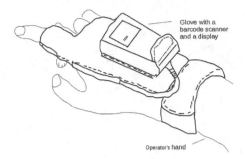

Figure 3.3 Gloves equipped with a bar-code scanner

3.3.4 BAGGAGE CHUTE LIFT TABLE

The chute lift table is positioned at the bottom of the baggage chute feed. Bags travel down the chute and feed onto the ball-table that forms the top surface of a lifting table (as shown in Figure 3.4). The table is raised to a height (≈ 0.9 m), which is considered as reasonable height to facilitate the pulling of a bag for most workers. It is used to lift bags over a buffer at the end of the chute. Thus, the bag is easy carried to the ULD for discharging at the appropriate level [19].

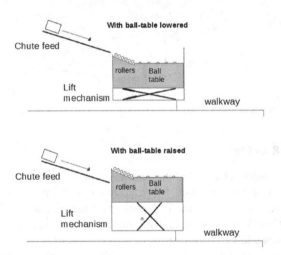

Figure 3.4 Baggage lift table

3.3.5 HORIZONTAL LOADING AIDS

Horizontal loading aids support baggage from below and are ideal next to carousels and conveyors. They guarantee a 100% lift of all bags into both open and closed

ULDs. Bags are loaded onto a platform that can be raised or lowered to the required height, which the handler moves into the correct position within the container.

With the platform kept at the receiving height, a handle is used to manoeuvre the equipment and the bag to the receiving container. Once under way, the worker may choose to raise or lower the platform (as appropriate) for the required bag discharge level in the container. Manoeuvring the device is bio-mechanically less stressful. The platform is then adjusted to the required bag discharge level in the container, and the bag is pushed off the platform into its position inside the container. The platform is then adjusted to the "travel height", and the device is pushed back to handle another bag and so-on [20].

Working postures when operating this system can be significantly better than manually handling the bags. There can be some requirements to bend the neck and back to get beneath the container roof, and when pushing bags off the platform at lower levels. This is likely to be a less frequent and less extreme requirement than when manually loading bags (as shown in Figure 3.5). In this case, the productivity is a bit less than the conventional manual handling of bags.

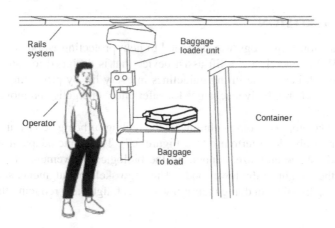

Figure 3.5 Baggage loader

3.3.6 BAGGAGE FORKLIFT

The Baggage Forklift is an efficient baggage loading and unloading system which can be used to eliminate and reduce the manual handling of OOG bags and enhances the processes quality and efficiency (see Figure 3.6). Indeed, it enables the airport to provide a safer working environment for baggage handlers and ensures that the OOG bags are handled gently and safely.

The Baggage Forklift is designed for easy operation using hand controls, the operator positions the unit in line with the baggage, slides the forks under the bag, and transfers it to a container [18].

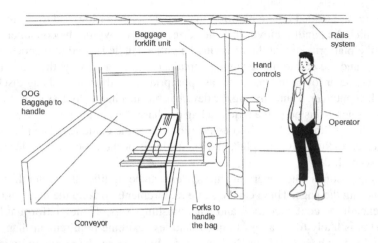

Figure 3.6   Baggage forklift

3.3.7   ERGO SKELETON

The Ergo Skeleton technology was designed to help protecting workers from back injuries [165]. In general, it is a lift assist device that is designed to automatically remind users to follow some lifting guidelines in every lift by promoting pivoting, knee bending and better body mechanics for safer, more energetic and more productive work [42].

The Ergo Skeleton technology eliminates the feeling of being "constrained", and allowing comfortable long-term use (see Figure 3.7). The frame adapts itself to the handler's body shape and size, supporting the biological movement of joints and transferring the weight to the lower body. The Ergo skeleton augments strength of the arms to help handlers in doing their job with less fatigue and reducing the risk of eventual injuries.

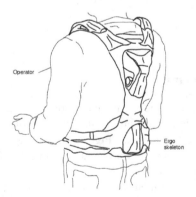

Figure 3.7   Ergo skeleton

3.3.8 SEMI-AUTOMATIC BAGGAGE LOADER

Semi-automatic loading aids are designed to use human interaction with the machine, and allow a certain freedom of control for operators. Such equipment requires minimal operator training and allows to increase productivity, minimise physical workload and the effective handling of peak loads. By using this equipment, the strain on operators is minimised.

The semi-automatic baggage loader is used for loading bags into containers with high efficiency and improved ergonomics. Using a simple joystick, the operator can control the flow of each bag into the container by moving the ramp to deliver the bag to its loading position. The operators load baggage into a container on an extendible conveyor (as shown in Figure 3.8). This conveyor can be moved in six directions (up, down, right, left, in, and out). To achieve all these rotations, the device is mounted on a moveable platform [21].

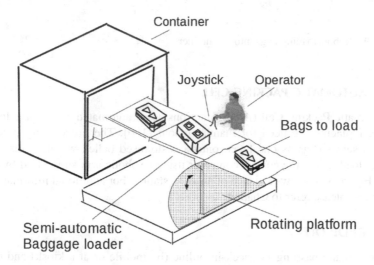

Figure 3.8 Semi-automatic baggage loader – sketch

3.3.9 ROBOT

The robots are fully automated loading aids, which use the eyes and ears of an operator without their physical input [161]. By using robotic technology, operators are upgraded to the position of a supervisor. Full automation is by far the most ergonomic and sustainable solution and requires the fewest operators.

The main feature here is the robot "gripper" which consists of a telescopic surface which holds the bag. And with the help of the 3D bin packing algorithm and the vision system, the robot can position the bag precisely above the pre-estimated spot inside the container (as shown in Figure 3.9). The robot gripper retracts its surface and the bag is placed down on the desired spot. This process duration is $\approx 10-20$ seconds, and repeats itself for every bag, until the container is full.

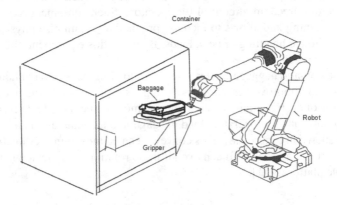

Figure 3.9   Robots loading bags into a container

3.3.10   AUTOMATIC PACKING CELL

The Automatic Packing Cell (APC) functions as an automated system for loading baggage, replacing the need for manual labour [180]. The process involves first, packing a set of bags as one layer, onto a customised pallet, which is then automatically loaded into the correct container using a robotic arm supervised by an IT system. Figure 3.10 shows a sketch of an APC station. For additional information on this process, please refer to Chapter 7.

3.3.11   DROP-OFF

Nowadays, many passengers check-in online (by mobile or at a kiosk) and obtain their boarding pass. Once they are in the airport hall, they use the Drop-off machines which look like a traditional check-in desk (see Figure 3.11). At these machines, the passenger hands over his bags and have them tagged then they are injected into the BHS. Finally, an optional receipt is printed for the passenger.

The baggage Drop-off machines deliver a wide range of benefits, including shorter passenger queues, reduced labour costs, etc. It is a way to free the check-in agents from lifting the bags and make savings for airports. Thus, the passengers become a kind of new handlers! Nevertheless, the problem is still the same underground, when it comes to operators to load quite heavy bags into the containers (nobody sees it!).

These drop-off points can also be installed remotely, at the hotel, at the metro network, at the car rental facility or at the parking of the airport, etc. In the near future, many airlines and airports will offer to their premium passengers the possibility to drop-off their bags at one of the above-cited points and proceed to the airport bag-free.

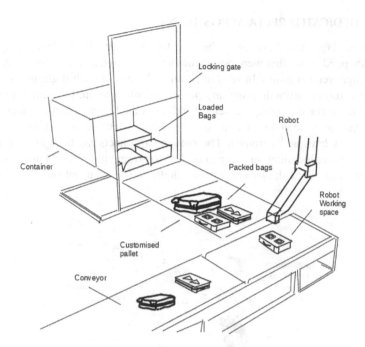

Figure 3.10 APC station – sketch

Figure 3.11 Drop-off station

3.3.12 DEDICATED RECLAIM POSITIONS

The dedicated reclaim positions can be used for reclaiming bags (the arriving ones), or even shopped goods that were bought on departures and stored in the storage until the passenger returns home. Instead of routing all bags of each flight to their corresponding carousels without using any passenger information, bags are temporarily stored until the corresponding passenger arrives in the baggage reclaim area [55].

The bags are presented to the passenger upon his identification, without the need to collects his bags at a carousel. The passenger collects his bags at a designated area. All bags are identified in order to be able to present the correct bags to the right passenger. Figure 3.12 shows a sketch of a dedicated reclaim station.

Figure 3.12   Dedicated reclaim station – sketch

3.4   CONTAINERS HANDLING

Inside the terminals, containers can be transported using Tug&Dolly, Roller conveyors, overhead conveyors, autonomous vehicles, etc.

3.4.1   ROLLER CONVEYORS

The Powered Roller conveyors are composed of rollers that are driven by motors. Thus, they offer huge health and safety benefits. Section 2.3.1 a short presentation of the powered roller conveyor technology.

3.4.2   CARGO HANDLING

Local cargo areas are used to store containers safely. In general, they are made of steel rollers and enclosed within a steel top plate. Due to the utilisation of the balls and rolls, the containers can be rotated easily, even in tight spaces allowing the personnel to push/pull the ULDs into any direction. Section 2.3.2 introduces the cargo handling and storage solution.

3.4.3 POWER TROLLEY SYSTEM

A Power Trolley System (PTS) comprises multiple trolleys (carriers) operating in a closed circuit, typically controlled by a management system [17]. This system involves the use of carriers in an overhead system to transport ULDs. Sections 2.3.3 and 6.4.1 provide an introduction to the power trolley system technology utilized for container transportation.

3.4.4 AUTOMATED GUIDED VEHICLE

The Automated Guided Vehicle (AGV) for ULDs was introduced in airports to transport ULDs. Thus, it frees up the personnel for higher-value tasks and also saves space. Within the terminal, the AGV connects several strategic handover positions (roller decks, conveyors, etc.) [83]. Sections 2.3.3 and 6.4.2 introduce the AGV technology which is used to transport containers at airports.

3.4.5 THE ULD TIPPER

The ULD Tipper is a mechanical unloading system used to unload automatically ULDs. It improves the working conditions of the handlers and has some benefits in terms of safety, since there will be less hands on the baggage. Used since many years in different industries (e.g., postal, warehouses, automotive, etc.). It took some time before it was first introduced to the airport industry [22].

The ULD Tipper allows to unload bags out of the ULD without any physical effort being required [182]. By lifting the ULD and rotating it 90°, the bags are sliding out in layers onto the conveyor belts (as shown in Figure 3.13). The bags can also be released by shaking the ULD. The baggage are then transported to their final destination (reclaim carousel or outbound area). These systems can empty a container in less than 5 minutes and can easily be adapted to different containers types.

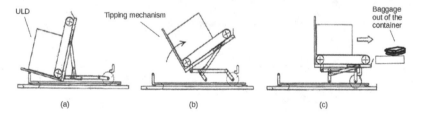

Figure 3.13 ULD tipping station

3.5 THE WAY FORWARD

The mechanical aid devices (e.g., manual, semi-automatic, and automatic) have been introduced to reduce the *manual* handling of baggage items and containers. These devices can significantly reduce the back-injury risk for handlers. Therefore, the utilisation of these devices should help to reduce the currently observed levels of discomfort and sickness absence particularly associated with back pain.

When these equipment are in use, the throughput rate is likely to be a bit low; however, these solutions require no physical lifting. Although it is accepted that there may still exist some minor manipulation of bags and containers, but this would be the same with the conventional manual processing.

The R&D projects in this field are still in progress. The operator (ground staff) acceptance of the loading aids is very important, and they would respond positively to any loading aid devices (with automatic scanning, etc.) only if it combines speed, ergonomics and value for money for the airport and the handlers.

4 Baggage Handling System Design

4.1 INTRODUCTION

An airport is made up of interdependent elements that can only be defined in relation to each other according to their place in this totality. The airport is a flow processing system of divers elements such aircrafts, passengers, baggage, freight, vehicles, etc. Overall, the role of an airport is to enable the transformation of travellers, baggage and freight into "batches" on board aircrafts in flight, and vice versa.

Usually, the number of passengers and bags per year are the main parameters that are used to define and design an airport. This is true from the macroscopic and the economical point of view. From the operational and the infrastructure point of view, what defines an airport is the number (average and maximum: peak values) of passengers and the number of bags on each flight over a given period of time (e.g., 1 hour). Thus, the size and capacity of airport infrastructure are defined by the maximum values, making the peak time the primary concern rather than the annual number of bags.

4.2 BAGGAGE HANDLING SYSTEM

A BHS can be defined by the following parameters:

- Daily number of flights: xx.
- Daily number of bags : xx.
- Annual number of bags : xx (in thousands or millions).
- Baggage Storage capacity : xx bags that can be stored.
- Peaks : around xx and yy.
- etc.

The airport capacity is the number of requests that can be processed during a given period of time by a set of equipment given the volume and desired quality of service offered by all the involved players. Thus, the capacity of the airport is one of the weakest link in the airport chain [170–172]. The capacity must therefore be the subject of a global analysis on all the links in this chain, namely: the terminal airspace, the runway system(s), the taxiways, the terminal (passenger handling), the baggage handling, storage, etc. The declared capacity is the maximum traffic flow that an airport is able to accept, taking into account all the elements of the airport chain as well as certain external constraints. This is the official capacity of an airport, which is necessarily less than or equal to the technical capacity [172].

DOI: 10.1201/9781003432920-4

To assess the capacity of an airport, several categories of parameters must be determined:

- the structural characteristics, which depend on the way in which the terminals are designed, laid out and equipped,
- the operating characteristics, which describe the operation of the terminal, the procedures for handling passengers and baggage, the mode of allocation of resources,
- the nature and the characteristics of the flows.
- etc.

4.3 DATA AND ASSUMPTIONS

In the next sections, some of the parameters that are used to analyse any Baggage Handling System (BHS) will be introduced.

4.3.1 AIRLINES AND HANDLERS

Let's assume an airport, where $1 .. h$ handling companies are hired by the airport for the baggage handling processes. Figure 4.1 shows an example of organisation.

A handling company often deals with $1 .. N_h$ airlines.

Each airline has $1 .. m$ flights per season.

An airline may have a code-share flight with other airlines.

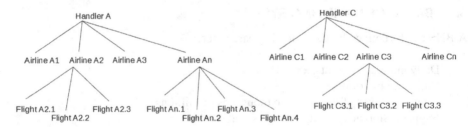

Figure 4.1   Airlines and handlers

4.3.2 FLIGHTS AND CATEGORISATIONS

Each flight is characterised by its airline, handler, departure time, number of ULDs, number of seats, number of bags, etc. (as shown in Figure 4.2).

A flight is also defined by the following parameters (Figure 4.3):

- Category : it can be Short Haul (SH), Long Haul (LH), Domestic (D).
- Load factor: the ratio of the number of passengers and the number of seats.
- BPP: number Bags Per Passenger
- etc.

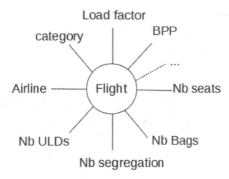

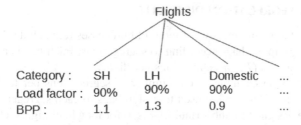

Figure 4.2 Flight characteristics

The number of bags per flight is given by the number of passengers per flight multiplied by BPP. Thus, if the load factor is 0.9, and the aircraft has 100 seats, the number of passengers is around 90.

Figure 4.3 Flight plan key parameters (example)

4.3.3 PASSENGER PRESENTATION TIME

The passenger presentation time is the temporal distribution of the probability that the passengers (and their baggage) stand in front the check-in desk of their corresponding flight. On the top of that we have the transfer baggage profile which is represented by the passengers upon their arrival to the airport via transfer flights. Figure 4.4 shows the check-in and the transfer baggage profiles.

Figure 4.4 shows the baggage profile (checked-in and transfer bags) of a given flight, which is the percentage of bags per a period of time inducted into the BHS.

The BHS capacity and its behaviour depends on the number of bags that can be handled at the same time. The load profile data is used to estimate the number of bags per period of time. The bag lists and the flight plans are used as a basis in the design phase of a BHS. The bag list is generated with segregation distribution according to the defined segregation rules of the flights as seen in §4.3.4.

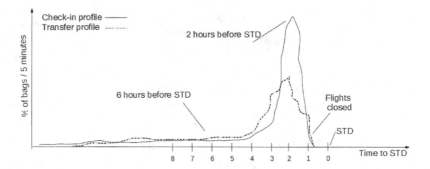

Figure 4.4 Check-in and transfer bags profiles

Each future critical situation is then translated into a real flight program over a representative period. This program, which describes the arrivals and departures of flights, includes all the information on destinations, timetables, filling rates, aircraft models, etc.

4.3.4 FLIGHT SEGREGATION DISTRIBUTION

For big airports, the main objective of the build processes (e.g., chute, MUPs, build cells, etc.) is to group the bags according to segregations. Each bag is assigned to a determined segregation (*1 ... n* segregations per flight).

Table 4.1 shows an example of the number of segregations per flight, whereas Table 4.2 shows the rules that are used to assign bags to each segregation. It shows that 3 MUPs are assigned to Short Haul flights, 6 for long haul flights, etc.

For further understanding, Table 4.3 shows an example of a Short Haul flight with 151 bags, 3 segregations. Let's assume that there are up to 35 bags per ULD. Based on the calculation, 6 ULDs are needed to load all the bags. Table 4.4 gives the number of bags per segregation as well as the number of ULDs per segregation.

Table 4.1
Segregations per flight category (example from big airports)

Flight category	Abbreviation	Nb segregations
Short Haul	SH	3
Long Haul	LH	6
Domestic	D	2
Exception	E	8
...	...	...

Table 4.2
Segregation distribution (example)

Nb segregations	% Seg 1	% Seg 2	% Seg 3	% Seg 4	% Seg 5	% Seg 6	% Seg 7	% Seg 8
1	100							
2	49	51						
3	29	40	31					
4	25	25	25	25				
5	10	19	20	21	30			
6	11	16	19	27	13	14		
7	10	15	17	25	12	13	8	
8	9	14	16	22	11	12	8	8

Table 4.3
Example of segregation distribution

Flight ID	Type	Nb Bags	Nb Segs	Nb ULDs
XXyyyy	SH	151	3	6

			Number of bags per Segregation				
Seg 1	Seg 2	Seg 3	Seg 4	Seg 5	Seg 6	Seg 7	Seg 8
44	60	47					

			Number of ULDs per Segregation				
Seg 1	Seg 2	Seg 3	Seg 4	Seg 5	Seg 6	Seg 7	Seg 8
2	2	2					

Table 4.4
Portion of OOG bags per flight (example)

Category	Summer	Winter
Out of Gauge	5%	10%
...	...	...

In general, especially for small airports, we have the following rules:

- 2 MUPs for Long Haul flights
- 1 MUP for Short Haul or Domestic flights

4.3.5 FLIGHT PLAN

The flight plan is used as a basis for the analysis of the baggage flow into a build area. Figure 4.5 shows an example of the number of flights per hour of the BHS of an airport. In this example, the average is around 10 flights/hour, and the maximum number of flights/hour is 35 (around 7:00 p.m.).

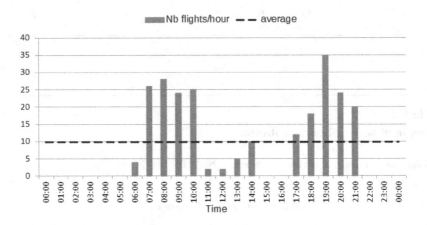

Figure 4.5 Number of flights per hour (example)

4.3.6 BAGGAGE FLOWS

Figure 4.6 shows the main parameters that define a baggage item, while Figure 4.7 shows the baggage flow in the system, which is the number of bags that pass through a designated position with the BHS (check-ins, screening machines, etc.). As stated earlier, the key element is not only the total number of bags per day or year but rather the number of bags per hour. This defines the peak and drives the whole capacity of the system.

Another parameter is the portion of the OOG bags per flight. This percentage can be different from one airport to another and from one season to another. Table 4.4 shows an example.

The oversized baggage is rarely a factor limiting the hourly capacity of the BHS. Nevertheless, their treatment must be studied during the design phase (risk of blocking certain modules): non-format check-in desks, explosives detection, appropriate sorting, transport systems, storage, etc.

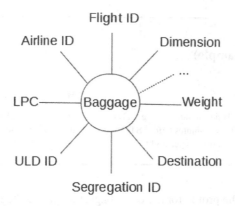

Figure 4.6 Baggage item characteristics

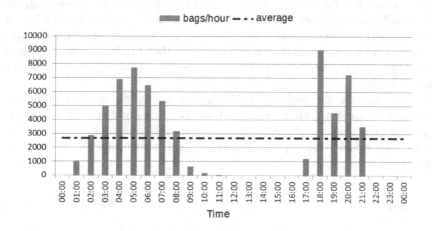

Figure 4.7 Baggage flow (example)

4.4 CONVENTIONAL FLIGHT BUILD

A build position (which can be a chute, lateral, etc.) is assigned to one flight/segregation for a limited period of time. Table 4.5 shows a build allocation example where the build positions are open and closed 150 (SH) (resp. 210 (LH)) minutes before STD[1]. Build positions are closed 30 (SH) (resp. 30 (LH)) minutes before STD. The number of allocated build positions per flight is 3 (SH) (resp. 6 (LH)). There is a dwell time of 15 minutes between a chute closing and reopening for another flight.

---

[1]STD: flight Standard Time of Departure.

Table 4.5
Build allocation (example)

	SH	LH	...
Build position Opening Time (minutes before STD)	150	210	...
Build position Closing Time (minutes before STD)	30	30	...
Number of allocated build position per flight	3	6	...

Figure 4.8 shows the profile for the early bags, the on-time bags and the late bags. No bags are allowed in the system before STD-x. More, no transfer bags are allowed in the system before that time.

In some airports, they make a first sorting by using the chutes. Thus, baggage items of different flights are dropped in the same chute (2 or 3 carts or ULDs for the same chute) and are then sorted with a thinner granularity by using the Baggage Reconciliation System (BRS). The handlers use the BRS handheld terminal to scan the bag tag, which enables the system to determine the specific ULD where the bag should be loaded.

In other airports, the number of MUPs is equal to the number of flights/ segregations (1 ULD = 1 MUP/chute) and, the gap between 2 consecutive alloca- tions to the same resource (chute, MUP, etc.) is set to 10 minutes per example. The early bags are stored in the Baggage Storage System (see Chapter 5). The late bags are sent to expedite chutes.

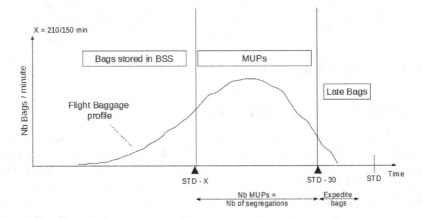

Figure 4.8 Conventional approach

4.5 MINIMUM CONNECTION TIME

The Minimum Connection Time (MCT) is the minimum time needed to transit at a given airport between two flights. The MCT is based on variables like the airport layout, the security checks, whether the connection is between international and/or domestic flights, etc.

In general, the MCT ranges from 30 minutes to 3 hours, depending on whether it is a domestic or an international flight. The MCT targets start from chocks on (inbound aircraft) to chocks off (outbound aircraft) and include ULD unloading and loading to the aircraft, etc. [170].

MCT= A + B + C , where

- A = (Aircraft arrival, engine run-down, doors open, first 2 containers unloaded and transferred to dolly ready to leave) + (Stand to baggage hall travel time) + (baggage hall unload time)
- B = Baggage system "In system time" (Departures or Transfers)
- C = (Baggage hall to stand travel time) + (baggage hall load time) + (unload container from dollies, load to aircraft hold, close doors)

4.6 IN SYSTEM TIME

The performance of a BHS is measured by the required time (called the In System Time "IST") for a bag that can be transported from one point to another within the system. Table 4.6 shows the different categories of IST [170].

4.7 PEAK SHAVING

Figure 4.9, shows the number of flights handled per hour in an airport. In the upper graphic, there are two peaks: the first one is around 08 a.m., while the second is around 07 p.m. In the bottom side of the graphic, there are no peaks, the number of flights remains stable for the two periods of time. The number of flights is the same in both configurations, and the only difference is the profile.

In the first graphic there are peaks twice a day, combined with a low activity. The second one, there are no peaks, but the maximum number of flights per hour is smaller. Therefore, we have the same pace the whole day. Sure, the second one is more than a dream, but it can be reached by using the peak shaving approach.

From a technical perspective, too much bags in the BHS will increase the risk of bottlenecks in the sorting machinery, which may result in delays and disruptions in the outbound baggage handling process. From the handler's point of view, the high workload peaks can lead to an unfair distribution of the work between the different handlers (colleagues). To avoid these problems, the handlers employ temporary workers that leads to increased labour costs [55].

The objective is three-fold: first, to minimise the number of flights operating concurrently in the same location; second, to decrease the number of bags checked-in

Table 4.6

Categories of IST

IST is the time that a bag takes	From (A)	To (B)
Originating Departures	Check-in induction onto collector belt (input) via hold baggage screening and flight allocation.	Presented (output) to the ramp operator either in a container (automated handling) or presented on a flight make up lateral ready for semi-manual handling and placement into a container ready for dispatch to the aircraft.
Transfers	Offload (input) from a container from a baggage handling operator in the baggage hall via hold baggage screening and flight allocation.	presented (output) to the ramp operator either in a container (automated handling) or presented on a flight make up lateral ready for semi-manual handling and placement into a container ready for dispatch to the aircraft.
Inter terminal	Offload (input-originating terminal) from a container, cart, dolly, or van from a baggage handling operator in the originating terminal baggage hall via cross-campus transfer conveying, hold baggage screening and flight allocation to the point at which it is.	Presented (output-destination terminal) to the ramp operator either in a container (automated handling) or presented on a flight make up lateral ready for semi-manual handling and placement into a container ready for dispatch to the aircraft.

early within the system; and finally, to evenly distribute the load and flow across different areas. These goals are crucial for operational efficiency as concentrating all activity in one area could result in a breakdown of the system.

The peak shaving is used (during the peak activity) to reduce the load on the system. The early bags, after their identification and screening, are stored temporarily in storage facilities. When the activity of the system decreases, the early bags are again released from the storage to be screened and sorted into their final destination.

Thus, the idea is to get the right baggage at the right time, at the right place at the right cost. Figure 4.10 shows the flight profile before and after applying the peak shaving. Below are listed some approaches to handle the peak shaving:

- High storage density using different storage techniques (see Chapter 5).
- Fast and random access to the stored bags.
- Semi-automated bags loading: saves labour and time (see Chapter 3).
- Baggage collection service (e.g., hotel, local drop-off, etc.).
- Workload balancing all day long.
- Decentralised baggage handling activities.
- Fast track for passengers and bags (rush hour).

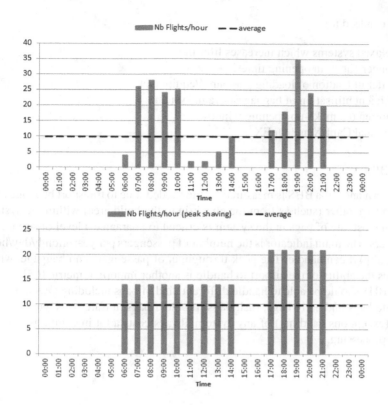

Figure 4.9 Standard flight profile (a) and flight peak shaving (b)

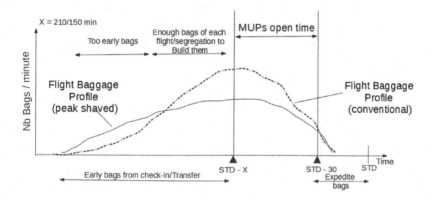

Figure 4.10 Peak shaving

This can lead to:

- Relaxed systems which increases lifetime.
- Longer check-in opening times.
- Better utilisation of loading personnel/equipment.
- Efficient utilisation of baggage storage system.
- Shorten the make-up opening window.
- Minimal CapEX and OpEX[2].

4.8 KPIS

The performance of a BHS is measured by the needed time to transport baggage from one area to another (such as from the check-in to the build area) within the system. More, the lost rate of bags in the system is crucial to guarantee a level of service for passengers. The main indicator is the number of passengers per year/month/day/hour, that the airport can handle. The peak throughput of passengers and baggage, which measures the ability of the airport to handle is another important metric [65].

The BHS should be able to handle different kind of bags including OOG, on-time bags, late bags and early bags, rush bags, problem bags, unsafe bags, etc. The fine tuning (exceptions handling) of any system always costs a lot in comparison to the normal processing.

---

[2]CapEx refers to a Capital expenditure while OpEx refers to an Operational expenditure

Part II

From Intra-Logistics to Airport BHS

5 Baggage Storage Systems

5.1 INTRODUCTION

The early bags (transfer or checked-in) that arrive before flight's baggage handling period has started[1] are directed to the baggage storage system (BSS) where they are stored until baggage sorting destination of their corresponding flight is opened. Later-on, bags are individually released from the storage and sent back to the sorting system. Then, the bags are loaded into the containers, which are later transported to the aircraft to be loaded into the cargo hold.

There are two fundamental characteristics of a BSS; the capacity and efficiency. The capacity, (i.e., the potential amount of baggage that can be stored), and its dynamic management (logic), that determines the efficiency of the storage system. The capacity of the storage system at some major airports is in general 10 times less than the total amount of bags handled per day. Efficiency, in turn, has a twofold definition. First, the methods that aim to use as much as possible the BSS capacity. At the opposite, those that smooth the output flow so that the BSS does not generate "peaks" when bags are released from the BSS.

A BSS management system is intended for the control of the BSS and the re-induction of the bags onto the sorting system when their time frame/flight indicates that their destinations will be accessible soon. When a bag arrives at the BSS area, it is sent to the desired storage location. The main aim is to use as efficiently as possible the available space in the BSS. The management system performs routine tasks of allocating physical storage (e.g., racks, totes, lanes, etc.), and re-organising the storage when necessary to smooth the output flows. After a determined storage time, the "locations" are emptied. The bags are tracked during this phase. The system can flush the buffers when conditions such as time frame are met.

5.2 THE TYPES OF BAGGAGE STORAGE SYSTEMS

The appropriate bag store solution for a specific area is mainly driven by two parameters: the number of bags to be stored and the average estimated "dwell time" of the bags, space, cost, etc. These parameters help to calculate the in- and output capacity required for the best BSS and thus optimise the flight build process.

There are many techniques to store early bags: on the floor, in buffer lanes, in cranes, etc. The conventional solutions use only one level of the building to store the bags. The recent ones, use the height of the building to store more bags. Figure 5.1 shows the evolution of the BSS solutions from storage on the floor, to sophisticated Automated Storage and Retrieval System (see section 5.2.6).

---

[1]In general, the flight's baggage handling period begins 1–3 hours before the scheduled departure time (STD) and ends 15–20 minutes before a flight's departure time.

DOI: 10.1201/9781003432920-5

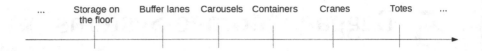

Figure 5.1 Evolution of BSS solutions

5.2.1 THE BAGS ON THE FLOOR SYSTEM

In small airports, the storage of early bags is done manually; thus, when a passenger has a layover of a couple of hours, the bags are placed in a manual storage area (see Figure 5.2). When it is time to load the bags, the staff needs to manually place the bag on the conveying system. This manual system consumes a tremendous amount of time, space, and money. Indeed, it is easy not only to store the bags on the floor but also easy to lose them. Often, they can be late.

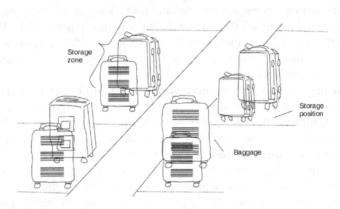

Figure 5.2 Bags stored on the floor

5.2.2 THE BAGS RACKING STORAGE

A more accurate solution is to use scanners to facilitate the manual storage and re-trieval of bags. The scanner is used to scan the bags label and firstly confirm that the bag is correctly routed and is to be held within the manual storage area.

The BBS store consists of racking with locations for bags on the basis of one location per bag (see Figure 5.3). The system maintains a map of the store and as such knows which locations are empty. The system instructs the operator to manually place a bag into an empty storage location via the scanner. Each storage location has an individual bar-code for identification. The operator is prompted to scan the storage location bar-code to confirm where the bag has been placed via the scanner.

Later-on the system instructs the operators to retrieve bags from storage via the scanner. The operator scans the storage location and the bag tag to confirm via the scanner that the correct bag is being retrieved. The operator then manually retrieves the bag and places it on a conveyor to feed the bag to the sorting machinery.

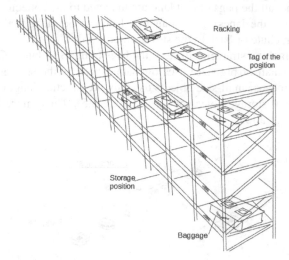

Figure 5.3 Bags racking storage

5.2.3 THE CONVEYOR LANES SYSTEM

A lane-based bag store is suitable in areas with limited height. The lanes hold bags from different flights but with a similar estimated/predicted departure time. The IT system controls the dynamic allocation of the lanes to the different dwell times. The storage system is composed of several parallel lanes to store the bags (see Figure 5.4). Each storage lane has an input queuing conveyor and an output – induction – queuing conveyor. Basically, bags are routed to a storage lane meeting a certain storage criteria. The storage conveyor is then moving, the bags one step forward, moving the first bag one step more (inching), etc.

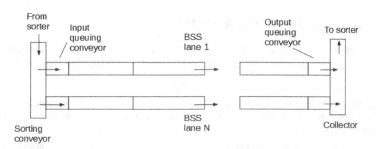

Figure 5.4 Baggage storage lanes

The amount of stored bags at a time is restricted by the space allocated to the lanes. After a determined storage time or when one or more bags should be retrieved[2] from a storage lane, all the bags of that lane are inducted to the collector (a conveyor lane that connect all the lanes). Each bag in a lane can be individually sent to its destination (sorter, chute or back to the storage area) [55].

There are two inching strategies to store the bags in the BSS lanes. Constant step means each bag (small or big size) uses the same space as the storage conveyor[3]. Constant interval means small-size objects use only their (actual length + a constant interval) segment of storage conveyor[4] (see Figure 5.5). Thus, more bags can be stored in this configuration.

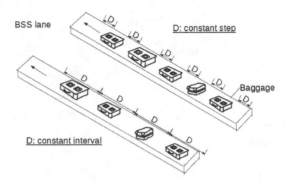

Figure 5.5   Constant step and constant interval BSS lane

In general, the bags will be stored upon their timing information. When a storage lane runs out of storage capacity, it might be reorganised into two separate storage lanes assigned to the succeeding time intervals (i.e., one storage lane covering 1 hour is split into two storage lanes, each covering half an hour). On the other hand, when additional storage capacity is required for a given criterion, an additional storage lane can be logically connected to the original one. Reorganisation of a storage lane is done by assigning new storage lane(s) and purging the bags of the storage lane to be organised, etc.

5.2.4   THE CAROUSEL STORAGE SYSTEM

In this configuration (see Figure 5.6), the bags are inducted into a carousel. The carousel has a certain capacity, which represents the number of bags that can be

---

[2]This requires to flush all the bags that are before this bag in the lane, which then requires to re-enter again the bags that were flushed. This approach can generate a huge load on the BSS system as well as on the sorting installation around it.

[3]For instance, to store 40 bags on one lane, a maximum 48 m (=40 × 1.2 m) conveyor length is needed.

[4]Suppose we have 40 bags with 0.7 m long and an interval of 0.3 m, then the total length will be 40 m = (40 *(0.7+0.3)).

buffered. Thus, these bags turn around on the carousel so far their final destination is not yet opened. Once opened, the bag is pushed out of the carousel and transported into its destination [178]. While this solution based on tracking baggage on the carousel is relatively simple to implement comparing to the other options, unfortunately, it is not energy efficient as the carousel keeps moving all the time.

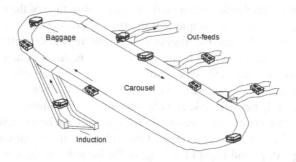

Figure 5.6   Carousel to store bags

5.2.5   TEMPORARY STORAGE IN CONTAINERS

One of the approaches is to store bags in containers (ULD, customised containers). Thus, the bags are loaded into containers[5] and stored on time or on flight basis. The bags are first sorted on the right lane, and they are then loaded manually or automatically into the containers. Once full, these containers are stored until their common release time is reached. Then, the containers are offloaded manually or automatically, and the bags are sent to the sorting area. The storage system is composed of loading and unloading stations as well as a storage area of containers (see Figure 5.7).

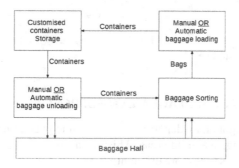

Figure 5.7   Utilisation of customised containers – processes

---

[5]To speed-up the loading processes, semi-automated loading devices can be used (see section 3.3). More, the tipping stations are used to offload these bags (see Chapter 3).

5.2.6 AUTOMATED STORAGE AND RETRIEVAL SYSTEM

Rather than storing bags in storage lanes or dynamic storage loops, the more efficient system is proving to be racking systems. They consume less physical space. Automated Storage and Retrieval System (AS/RS) (see Figure 5.8) is another method of storing bags and is based on rack positions served by stacker cranes – a technology that originates from warehouse automation. A strong advantage of the racking-based BSS is the "random storage" principle. All the storage positions can be individually accessed by cranes or shuttles, the bags can be stored anywhere until their final retrieval. This is especially practical in the case of batch building, when – at the time of storing – it is not known which batch build the bag will be part of (see Chapter 8).

The warehouses (cranes, shuttles, totes, etc.) can hold a thousand of bags. The system is considerably more flexible and efficient than other solutions because the lift performs the vertical movements, and the shuttles perform the horizontal movements. The baggage can be stored either as "raw" (loose) bags or in tubs (totes). This latter has the advantage of providing full tracking of the baggage in the lane; however, it requires more space due to the size of the tubs. Nevertheless, the primary advantage of using the tubs is their ease yo automation in handling. Indeed, the system moves tubs that all have the same dimensions, rather than having to push and to grab bags of various shapes and dimensions.

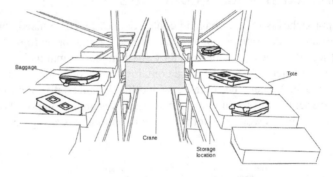

Figure 5.8 AS/RS bag storage

5.2.7 MAKE BAGGAGE STORAGE MULTI-PURPOSE

The fact that many flights may depart at the same time puts a huge pressure on the restricted spaces of baggage build areas. This means airports have to work out how to make their storage systems more efficient.

Many airports have made their existing early baggage storage (EBS) systems ready for the future. Thus, in combination with the batch building concept, transfer bags, rerouted bags, the reclaim on-demand bags, etc. the airports can turn these systems into multi-use storage systems. These systems will surely boost the baggage handling processes and improve the use of valuable airport building space and resources [14].

6 Individual Carrier Systems

6.1 INTRODUCTION

Can the Baggage Handling System (BHS) transport baggage from one point to another as fast as the travellers can do? This is a simple way to measure the performance of a BHS. Thus, if the bags are transported slowly: the passengers will be frustrated waiting for their bags, or bags failing to make connecting flights on time. Thus, the time required for a baggage to be transported from the check-in area to its destination is determined by the speed at which passengers can make the same journey to the boarding gate. In some airports, it might only be a short walk to the boarding gate, while in others, passengers might have to take a bus, train or even walk long distances.

Airports desire fast baggage delivery and real-time and accurate tracking of their whereabouts, fast connection time of transfer bags, and the ability to serve remote terminals efficiently. The BHS must also be able to handle a large variety of different bag types, random access bag storage, etc. Conventional technology (such conveyors) cannot provide all of these services, but autonomous vehicles can.

The Baggage Individual Carrier System (BICS) technology is based on carriers used to transport baggage. In a BICS unit, each bag is loaded individually into a carrier and stays within the carrier, from check-in desks till its final destination. There are basically three main types of BICS (see Figure 6.1):

(a) Automated Guided Vehicles (AGV): it is a metal cart (with wheels on the bottom) and a plastic tub on top.
(b) Tote: A simple cart provided by motors situated around the track.
(c) Destination Coded Vehicle (DCV): A cart being powered by an on-board linear motor and running on a simple track.

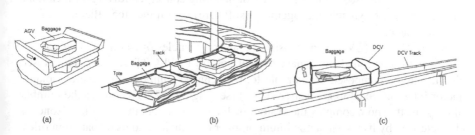

Figure 6.1 Baggage Individual Carrier System – concepts

A BHS is like a city's roads and the conveyors are like the local roads, while the BICS tracks (physical or virtual) are like the highway and the bag (on the BICS) is like the car [24]. If a BICS track is blocked (a traffic jam), the vehicles can be routed around the blockage area. The bags start and end their journey on conveyors (as you start your drive on local roads), moving to the BICS track to make longer journeys, such as from one terminal to another or gate to gate. The BICSs almost never stop, just as there are no stop lights on a highway (as far as there are no blockages) [125].

6.2   THE BAGGAGE INDIVIDUAL CARRIER SYSTEM

In the 1980s, the choice for airports expanded from conventional belt conveyors with pushers to tilting tray sorters. The early 1990s saw the introduction of the modern DCV. The DCV's objective was to place a baggage item on a cart (on a track) and deliver the bag quickly to a gate or to an appropriate loading area. Today's airports are getting bigger and as a result, the travelled distance by baggage inside the BHS is increasing. At the same time, check-in and transfer times need to be reduced in order to better serve passengers. Thus, long distances can be one of the justification for DCV technology [108]. The DCV systems are appropriate when the shape of the building of the BHS area does not allow easy installation of traditional conveyors (a lot of pillars, no space for long straight lines, no space for a sorter, etc.)

A Tote system can be one of the best solutions to airports with one or more terminals. It can handle many drop points over long distances. Cart-based systems fit well with smaller terminals that need high-speed transport capacity. The solution is suitable for medium and high-capacity sorting systems where medium-speed transport and a limited number of destinations are needed [14].

The AGV is foreseen to efficiently transport bags that enter the BHS premises at check-in desks or transfer inbound to their final sorting destination passing through EDS, MCS, etc. It can also be used to transfer bags from one sorting machinery to another, etc. [48, 175]. They are used to handle bags by exception such as oversize bags, rush bags, transfer bags and inspection bags, or as a standalone end-to-end baggage handling system in place of a conveyor-based system, etc.

There are two fundamental characteristics of BICS; its capacity and its efficiency. Its capacity, i.e., the potential amount of bags it may handle, depends on its length and the number of used vehicles, its way of handling convergent/divergent locations, etc. Secondly, its dynamic management (software) determines the efficiency of the capacity usage.

An AGV or a DCV system consists of several vehicles operating in a closed circuit, usually controlled by a management system (a PLC/IT system). The closed circuit may be composed of a simple loop or can be composed of several isolated and/or interconnected loops (sub-systems) (see Figure 6.2). The bigger the number of loops is, the more complex the system is. Each sub-system that is called "zone" is characterised by its layout, a maximum number of vehicles and its dispatching rules. For the Tote system, the carts are stacked and moved back to the induction area.

A BICS can be used when the distance from the main terminal to their destination is quite long. The vehicle can travel many times faster than a conveyor – almost

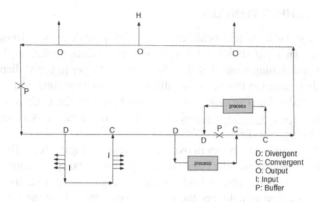

Figure 6.2 BICS system

8-10[1] m/s. Once, an empty vehicle is in front of the loading conveyor, the bag leaves the end of the conveyor belt and slides into the tub. Unloading a BICS is a similar process. Dynamic loading is a process where vehicles do not stop when loading or unloading a baggage, they just slow down. In contrast, vehicles of static loading systems must stop to load or unload baggage.

A mission *(task/job)* is an action to be carried out by a vehicle. Each vehicle is characterised by its status (empty, full), mission (planned, in progress, accomplished, etc.), task (load, unload, etc.) an origin and destination (check-in, transfer inbound, etc.), path (list of temporary destinations).

6.3 THE BICS MANAGEMENT SYSTEM

The BICS Management System (BMS) is the PLC/IT system that is used to manage the vehicles and the junctions. Its purpose is to manage vehicles (dispatching, routing, buffering, etc.) of an area of the system. The BMS performs routine tasks of allocating a vehicle to a bag and reorganise the number of vehicles attributed to each zone (sub-system) when necessary to smooth the global output flows. Thus, the BMS accommodates mainly two core system functions. The first one is to assign missions to vehicles to transport bags from one location to another. The second one is to organise the zones (assign vehicles to zones to handle changes in flow) [199]. The routing and the dispatching problems are introduced in Chapter 14.

One of the major challenges in BICS systems is the necessity to manage the flow of empty carriers, which consumes a significant part of the "bandwidth" of the installation. Another challenge, linked to the first one, is to correctly manage empty carrier storage zones, in order to ensure always having an empty carrier when needed.

---

[1] It is four to five times faster than the 1–2 m/s belt conveyors.

6.3.1 DISPATCHING VEHICLES

Assigning jobs to vehicles and vehicles to jobs (dispatching) is a decision rule to select a vehicle, an input station (pick-up) and drop location (delivery). There are two groups of dispatching strategies. The first one is <u>station-initiated</u> dispatching, in which a vehicle is selected from a set of idle vehicles to be assigned to the next task in the queue of tasks. Such strategies are selected based on the BICS system type in use. The second type is <u>vehicle-initiated</u> dispatching, in which a task is selected from a set of requests for a service.

The dispatching rules can be set in two categories (see Figure 6.3). The first one is to select a vehicle among the idle ones and assign it to a transport request generated somewhere in the system. The second one is to select an input station among the stations requiring a free vehicle. For the first category, the rule of the nearest vehicle is used to satisfy as quick as possible requests for a vehicle (distance criterion). For the second category, among others the rule of the maximum outgoing tail size rule can be used. Indeed, the station presenting the full buffer is selected to avoid exceeding the capacity of the buffer (buffer criterion).

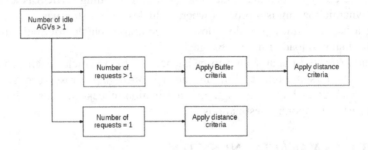

Figure 6.3 Dispatching vehicles

Other rules can be added to the two preceding ones, for instance, the rule reducing the waiting time, where the goal is to increase the flow. Of course, one can consider the case where a dispatched vehicle interrupts its mission to take another one (Figure 6.4). There are simple and pre-emptive dispatching rules, and with disparate objectives. Such strategies include:

- First coming first served
- Simple carousel
- Short-sighted starting
- Missing station without reassignment
- Missing station with reassignment
- Pure cycle without reassignment
- Pure cycle with reassignment
- etc.

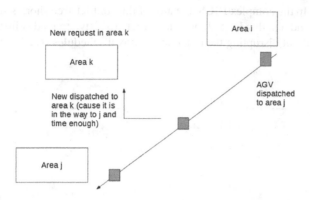

Figure 6.4 Re-dispatching rules

The BMS uses the topology of the system for dispatching the vehicles. The length (total distance), as well as the desired throughput of the system, defines the maximum number of needed vehicles. The topology is a mathematical (artificial) tool that is used to describe the architecture of the system to be managed. Simply, it is characterised by its segments between the different locations and the nodes.

To find the shortest path in a graph (topology of a system), the Dijkstra algorithm can be used [43]. The goal of the Dijkstra's algorithm is to find the shortest path between a given node, and the other nodes of the graph. Given a weighted (positively) graph and a starting node, the algorithm determines the shortest path as well as the distance from the source to all the destinations in the graph (see Figure 6.5).

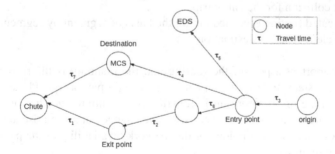

Figure 6.5 Topology – example

6.3.2 THE VEHICLE ROUTING

Once the dispatching problem is solved, the BMS has to handle the routing and the planning of the vehicles. The routing of vehicles consists of selecting the best way to reach its destination. The planning of vehicles is done in real time to avoid collisions and obstacles. In the complex BICS, the aim of the conflict-free shortest-time routing problem is to find a path which allows the vehicle to arrive to its destination as early as possible without disrupting the other active travel schedules (see Figure 6.6).

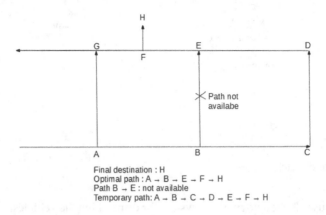

Final destination : H
Optimal path : A → B → E → F → H
Path B → E : not available
Temporary path: A → B → C → D → E → F → H

Figure 6.6   Routing of a BICS

One of the basic ideas is to maintain a list of free time windows or time intervals at each node and to route the vehicles through the free time windows. The objective is to spread up the traffic in the network to reduce traffic jams. Thus, the routing and the planning can be executed in two ways:

- The complete routing finds the path for the vehicle, i.e., from its current location towards the destination. In this case, the planning must assure a trip without collision for the entire trip.
- The repeated routing is carried out on the basis of segment by segment until the vehicle reaches its destination.

Another important aspect of the routing and the planning is the response time of the control system. Indeed, it is essential to select a path as quickly as possible. The response time of an algorithm is function of its simplicity. The robustness is defined as the algorithm capacity to consider the changes that can arise at any time. These changes concern the topology of the network (availability of the paths) or the changes in the requests.

The routing algorithms can be classified in two categories: non-adaptive (or static) algorithms and adaptive ones. The static algorithms do not consider the current traffic, nor the topology of the network, to select a trajectory. The path between nodes "i" and "j" is calculated in advance. These algorithms are easy to use due to their static nature; they have a very short response time. Nevertheless, they cannot adapt to the changes that can arise in the system. If the traffic is predictable, these algorithms suit very well. While, if the traffic varies, they are inefficient.

With the adaptive algorithms, the routing task is based on the current system's state. Thus, it can be:

- Centralised: Routing is realised by the BMS. The decisions are taken based on the information available on the entire network. Thus, the complete path, taking a vehicle from its current location towards its destination, is selected, the vehicle follows the sequence of nodes.
- Local: It is a decentralised decision process where each vehicle reaching a node will select the next node to follow. The next node is selected per the distance between itself and the neighbours and the current local flow of vehicles. The node presenting the shortest time trip is selected.
- Distributed: It is a combination of centralised and local algorithms. A vehicle arriving to a node will select the next node to take, but the decision is taken per local and global network information[2].

The centralised method presents a response time proportional to the dimension of the network and the number of vehicles to control. Also, it can select the optimum solution, i.e., the short path taking to destination. The local method is simpler and presents a short response time. It is robust to select the next node to take, but it can move away from the optimum path to reach a destination. The distributed method is simpler than the centralised method, but it does not guarantee the optimum.

6.4   THE ULD TRANSPORT SYSTEMS

The transport of ULDs from/to the baggage build areas to the stands is critical for meeting minimum connection times of transfer bags. Figure 6.7 shows the evolution of the ULDs transport systems. From manual, to motorised Tug&Dolly, to Overhead conveyors, Autonomous vehicles, etc. This field is still in continuous developments.

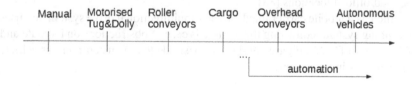

Figure 6.7   Evolution of ULD transport systems

6.4.1   THE POWER TROLLEY SYSTEM

A Power Trolley System (PTS) consists of several trolleys (individual carriers) operating in a closed circuit usually controlled by a management system. The closed

---

[2]The problem is similar to the one solved by an intelligent car GPS application, which adapts the route of each vehicle based on its knowledge of the overall traffic.

circuit may be composed of a simple loop or can be composed of several isolated or interconnected loops (sub-systems). Switches (divergent or convergent) are used to move between the different areas (loops). The carrier speed on straight lines can reach 2 m/s, and 0.5 m/s in curves. Figure 6.8 shows a carrier of PTS transporting a ULD. The routing and dispatching are similar to the principles used for the BICS transporting bags.

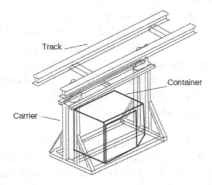

Figure 6.8   PTS system

6.4.2   THE AUTONOMOUS VEHICLES FOR ULDS

The shuttle bus self-driving vehicles in the airport, transports passengers from the car parking lot to the terminal. These vehicles have no steering wheel and are capable of travelling freely along a designated route. Similarly, there are autonomous vehicles that transport ULDs inside the airport. Each vehicle is capable of transporting autonomously the ULDs from one location to another and interface directly with the BHS the load/offload locations [83].

The vehicle is a mobile carrier thanks to an omnidirectional drive system and uses a laser scanning system redirecting the vehicle is case of obstructions on its route and can reach a speed of 3 m/s. Figure 6.9 shows a basic sketch of an autonomous vehicle to transport ULDs in the airport.

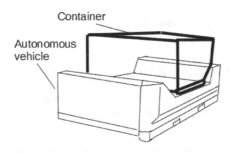

Figure 6.9   Autonomous vehicle – sketch

7 The Automatic Packing Cell

7.1 INTRODUCTION

In the last decades, the number of passengers and bags have increased in airports. Ground handling agents pack these bags into containers manually, and this requires large labour forces and space. The baggage handling is a labour intensive work, and physically very demanding. The work-related injuries cost money to the airlines.

The most important Key Performance Indicator (KPI) in the BHS field is the average cost per handled bag, and this depends on the used equipment, number of mishandled/delayed bags, OPEX, CAPEX, etc. The manual handling of bags presents a high risk of injury to the handlers carrying these tasks (see Chapter 3). These risks can be reduced significantly by considering the job content and using automation.

The Automation is a process that use machines to automatically perform tasks that can be performed by humans. At airports, robots may save handlers from lifting a dozen of tones every day. They can handle any type of bags, anything up to 50 kg.

The investment in one automated unit can save insurance costs and bring added safety and operational benefits, while fitting in a smaller space. Thus, the conventional conveyors transport bags into the loading area, and the robot replaces the manual labour to load the containers automatically and frees staff members for more complex tasks.

7.2 THE BAGGAGE LOADING APPROACHES

At many airports, most of the baggage-loading operations are still performed by handlers, manually loading bags into containers, and so-on. Bags are loaded into a carts, then they are transported to the aircraft, where they are individually loaded onto the cargo hold of the aircraft. The process is a time-consuming, expensive, and laborious process.

Different approaches can be used to load the bags into containers (see Figure 7.1):

- Manual handling (conventional).
- Handling aid (MHAD : bag lifter, hook, etc.).
- Semi-automatic handling (Semi-automatic baggage loader, etc.).
- Automatic handling.

Figure 7.2 shows the impact of automation on the ergonomics and the process time. The automation improves the ergonomics and reduces the process time, and the main exception is the manual handling aid devices, where the process may be slower comparing to the purely manual one.

DOI: 10.1201/9781003432920-7

Figure 7.1 Evolution of baggage handling techniques

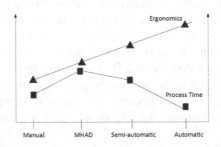

Figure 7.2 Impact of automation on ergonomics and process time

7.3 THE ROBOTS

The Robot solutions for the baggage handling use a 3D baggage assessment system that captures the size, the volume and the consistency of the bags (see Figure 7.3). This information is used by the IT system that calculates the best loading position of a given bag. A vision system continuously checks how the bags packed inside the container in order to re-calculate the optimal loading sequence. This approach leads to an average cycle time of 20 seconds per bag, means 12 minutes per ULD (35 bags per ULD in total) in case of a continuous flow of bags into the robot cell.

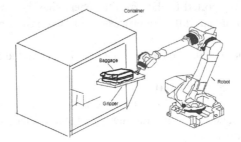

Figure 7.3 BHS robot

There are many robotic solutions to load bags into the containers. These solutions bring added work safety, savings to manpower and operational costs to the ground handler. Airports can save valuable floor space as these solutions require much less space than the conventional manual packing process.

Nevertheless, robots are still not totally acceptable if compared to the conventional manual handling solutions. The savings generated in OPerating EXpenses (OPEX) wise, are lost in the CAPital EXpenditures (CAPEX) wise, since these solutions require a big baggage buffer and experience a slow loading process time.

7.4 THE AUTOMATIC PACKING CELL

The Automatic Packing Cell (APC), is also an automatic baggage loading system equipped with the artificial intelligence (2D Bin Packing) algorithms which replaces the manual labour [180]. All bags are first stored in a BSS, and then released into the APC area through conveyors (Batch build concept: see Chapter 8). A robot with the help of its robotic arm (supervised by an IT system) automatically loads a set of containers (1 ... n) as and when they are required. An APC is composed of robot (a loading unit), a working place, a customised pallet (where to pack the bags), pallet storage and a transport system (see Figure 7.4).

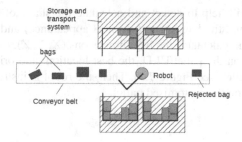

Figure 7.4 Packing cell station – architecture (top view)

With the help of a 3D vision system, bags are analysed and their sequence or location is planned before being handled. The system collects information on each bag's weight, size, shape, etc. (see section 7.4.1).

Several different approaches to loading bags into the customised pallet can be used (e.g., grippers, vacuum unit, pick and place, pushers, etc.). The robot packs bags on a customised pallet (full of bags : one layer) that is being loaded into the corresponding container. The customised pallet is only used to load bags, and it is pulled out after every loading step from the container.

Figure 7.5 illustrates the contrast between the robot and the APC approach for loading bags into a container. With the robot, the bags are loaded one by one into the container. In the case of the APC approach, a single layer of bags is first prepared on a customised pallet and then loaded into the container in one step. In the diagram, the bags are represented by small boxes, while the customised pallets are depicted as large boxes.

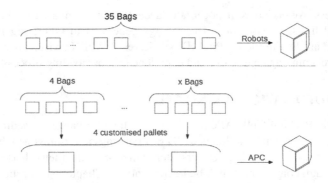

Figure 7.5   Baggage loading approaches

7.4.1   THE VISION SYSTEM

The vision system will help to detect and calculate the shape of the bags. Indeed, the camera will capture the bag image (physical appearance) and send it to the IT system. The IT system calculates the bag dimensions (X, Y, Z), the orientation (R), the remaining space on the pallet/ULD, the best location, and orientation of a bag on the customised pallet (see Figure 7.6). Then, the robot is instructed on how and where to put the bag on the customised pallet.

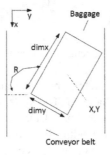

Figure 7.6   Baggage dimensions and orientation (top view)

7.4.2   THE 2D BIN PACKING

The bags in the APC area are delivered in a random manner and are characterised by their ULD, layer ID, dimensions, and position (x, y, z). The 2D bin packing algorithm is used to pack the bags on the pallet (see Chapter 14). Additionally, the 2D/3D bin packing is used to pack the bags on the customised pallet as well as inside the container. The robot then transfers the bag into the customised pallet. The bag's position on the pallet depends on its dimensions as well as on the space available on the pallet (see Figure 7.7)).

The Big Data and Deep Learning concepts can be used to learn from previous packing of bags. Indeed, it is the shapes of the bags and their positions inside the container that lead to good stacking of bags. Thus, if the algorithm can use the dimensions of the current bag to pack, the space available on the pallet and a database of previous staking, this can help to choose the right place for the bag at hand, and lead to improve the baggage stacking.

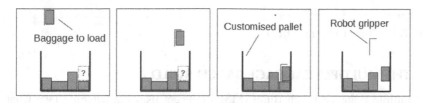

Figure 7.7 2D Bin Packing – customised pallet

7.4.3 THE CUSTOMISED PALLET

The customised pallet is a mechanical system that is used to load a set of bags into the ULD. The moving part of the customised pallet is used to push bags at the inclined side of the container (see Figure 7.8). The bags are moved in a smooth manner in order to keep the layer as balanced as possible.

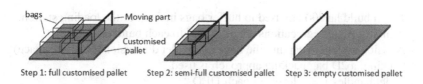

Figure 7.8 Pushing the bags inside the ULD – customised pallet

7.4.4 THE LIFTING TABLE

The ULD is positioned on the lifting table, while a customised pallet is set at a fixed height. A Robot packs bags on a customised pallet, which is then loaded (layer per layer) into the ULD. To load bags into the different levels of the ULD, a lifting table is utilised. Once a layer is full of bags, then the lifting table is lowered. Thus, the new layer full of bags is loaded into the ULD, etc. Figure 7.9 depicts the principle of the lifting table.

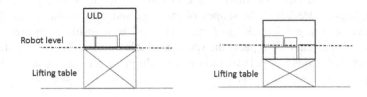

Figure 7.9   Lifting table of the ULD

7.5   THE MULTIPLE BAGGAGE BATCH BUILD

The Multiple Baggage Batch Build (M3B) concept uses the batching (see Chapter 8) and the automatic baggage packing cells. This approach gives the handlers the possibility to deal with the bags in an ergonomic and cost-efficient way. Besides its flexibility, it may reduce workers stress, ensuring best practices and cost savings, and there may be a considerable amount of space saved too.

The M3B concept may result in more productivity per square metre building and per handling resource, while significantly improving the working conditions, process quality and bag security. The M3B basic idea is to allocate an APC to more than one build (flight/segregation) per build time. The build time is the time needed to load (2, 4, 6, etc.) containers at the same cell. Thus, the resource is shared and the BSS size is reduced. The M3B building blocks can be summarised as follows:

- 1... n build lane(s) are used to buffer bags in front of the loading station.
- A mix of flight/segregations are allocated to each build lane.
- A vision system to capture the bags and pallet dimensions (for free space)
- Packing: a 2D bag packing approach is used.
- Pallet handles 1 ... n levels of each ULD.
- Robot loads bags into the pallets.
- Pallets are then loaded into the ULDs.

The M3B Baggage Factory build is based on the following steps (see Figure 7.10):

1. On time bags are sent directly to the APC.
2. Early bags are packed in ULDs (in the APC area) and buffered in the BSS.
3. The APC is used to load automatically (ULDs).
4. Batch build management system allocate build lanes and APCs to flights/segregations, etc.
5. At a determined time, bags (batch) of a flight/segregation are released from the BSS into build lanes.
6. The ULD Tipper is used to unload automatically the ULDs.
7. Bags are inducted to the sorting area.
8. Build lanes are used to store bags per flight/segregation (not far from their corresponding APC).

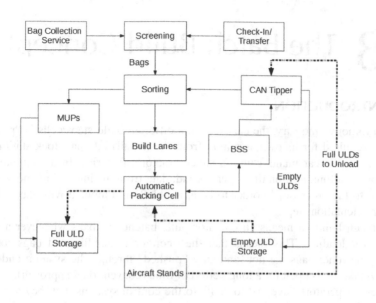

Figure 7.10 M3B Baggage factory – general flow

9. One build lane can be dedicated to more than one flight/segregation.
10. At a determined time, batch of bags are released from the build lane into their corresponding APC.
11. Robot loads packs bags into the pallet.
12. Full pallets are loaded into the ULD (on-time bags) or Customised CANs (early bags).
13. ULDs are sent to full ULD storage or to the aircraft
14. Customised CANs (full of early bags) are sent to the BSS for storage.

8 The Batch Build Concept

8.1 INTRODUCTION

In the automotive industry, the car along the different build steps/cells is "pulling" the parts required for its own assembly from the (small) internal stock stations, or directly from the car manufacturer's subcontractor's factories. In a computer, the processor is pulling, million times per second, data from the internal memory or the hard disk to the processor in order to be assembled or computed with the result of the prior calculation step.

The batch building means an operator pulls batches of bags whenever a batch is ready for loading. This means that the operator is handling all bags for one flight at one time; bags are not randomly "pushed" through the system; and none of chute/make-up is open for a long period of time (conventional approach).

Make-up operators have just to "tell" to the control systems that they are ready to receive bags on the chute/make-up where they are standing. The control system releases the bags from the bags store and "Opens the Chute/Make-up position" as long as the operator has not closed the batch [14].

8.2 THE MAKE-UP POSITIONS ALLOCATION (CONVENTIONAL)

The Make-Up Positions (MUPs) allocation system allocates the final destinations (chutes, carousels, MUPs, etc.) for outbound flights based on the strategy defined by the airport. These rules are used by the SAC system to sort the bags into their destinations. This allocation is built using the generic flight data (per season for example) and from the data provided by the Flight Information Management System (FIMS)[1]. Figure 8.1 shows an example of the conventional MUPs allocation of the flights.

8.3 PUSH VS. PULL APPROACHES

Like the conventional manufacturing, a push system, produces products in quantities that the company can do, and then gets it out to its customers. This result in large inventories at the end of the line. A pull system only produces what a customer has asked for. Thus, the push is "make to stock", while the pull is "make to order" [99].

8.3.1 PUSH APPROACH

Most BHSs are "push system"; passengers check-in or drop-off their bags and these are then pushed towards their final destination. Early bags are held in baggage storage

---

[1]In real-time with updates from FIMS by considering the modifications concerning both the flights already allocated and the new flights to allocate (see Chapter 9)

DOI: 10.1201/9781003432920-8

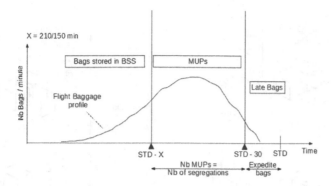

Figure 8.1   Conventional approach

until they can be handled with the rest of the on-time bags. Late bags have to be expedited outside the normal system to make sure they arrive on time to be loaded, into the corresponding aircraft, before departure.

Figure 8.2(a), shows a representation of the *Push* approach, a resource is assigned to a given flight-segregation on time basis (a static planning). Empty ULDs are also brought to the building on time basis. A resource is assigned to a flight/segregation even if there are not enough corresponding bags available and/or their corresponding ULDs are not yet in the building. This results in the fact that the resources are not fully utilised to their capacity.

8.3.2   PULL APPROACH

Warehouses and BHSs, seem to be completely different, but when you look closely, the similarities are obvious. Both systems receive, sort, store and forward items (parcels or bags) onto their final destination. Warehouses are in general "pull systems"; they are triggered when an order is submitted resulting in its items being (picked) pulled from storage, collected into orders and then batched by location into loads ready to be placed on a vehicle to take them to their next destination. In the parcel business, the aim is to pull everything together beforehand and load it as quickly as possible onto the truck to minimise its waiting time at the docks [129].

Figure 8.2(b), shows a representation of the *Pull* approach which is one of the basics of the lean management. A resource is assigned to a given flight/segregation based on the available bags and the corresponding ULDs. The bags are stored in buffers and are released in batches only when there are enough bags of the same flight/segregation, their corresponding ULD ready in the building, and a resource has been allocated to that flight/segregation. This can result in a higher utilisation of the resources, which will reduce the cost per baggage handling.

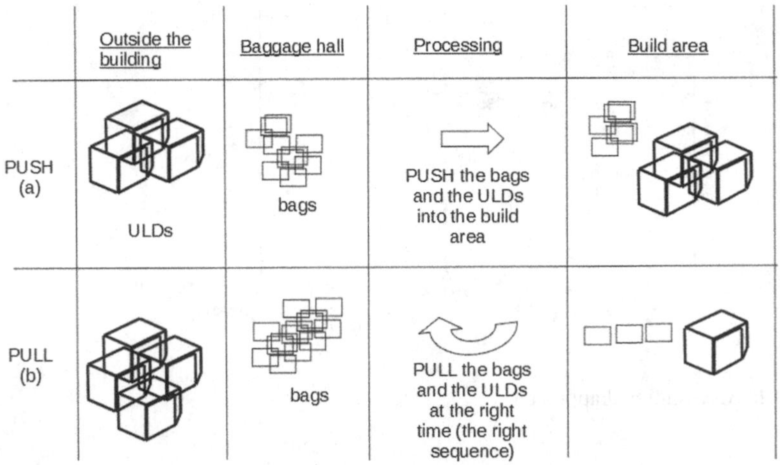

	Outside the building	Baggage hall	Processing	Build area
PUSH (a)	ULDs	bags	PUSH the bags and the ULDs into the build area	
PULL (b)		bags	PULL the bags and the ULDs at the right time (the right sequence)	

Figure 8.2 Push approach (a); Pull approach (b)

8.4 AUTOMATED BATCH BUILD

A batch build process (loading of bags in groups) enables a more efficient use of people and building space, which in turn increases the productivity of the BHS. Automated Baggage Batch Build can be defined as a process to load the bags into the containers in batches which are scheduled and managed automatically. The batch build concept can also be defined as a fact *to execute the right job at the right place, at the right time, where the required resources are available, as efficiently as possible, within a batch window.* The workload is balanced as possible among the different resources. The goal is to synchronise the bags (flight/segregation) and their corresponding ULD at the build area [36, 96]. Figure 8.3 summarises the batch build concept.

The main resources of the batch build concept are presented below. Figure 8.4 depicts the flow between these resources:

1. Baggage storage system
2. Build lanes
3. Empty/full ULD buffer
4. Build cell

8.4.1 THE BAGGAGE STORAGE SYSTEM

The baggage storage management system is intended to manage the bags and ensure their proper delivery to their respective destinations as soon as the build cell for their flights/segregation/time frame is about to be opened.

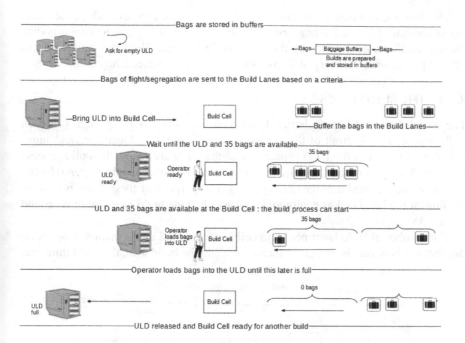

Figure 8.3 Automated batch build

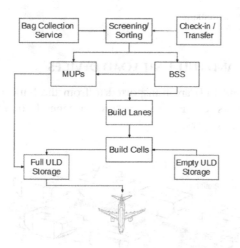

Figure 8.4 Batch build concept

The dynamic bag storage system is based on a modular approach where the quantity of modules and the feeding systems allow for high flow rates to accommodate the dynamic operation driven by the batch build process. The batch build concept asks a "just-in-time" delivery of the bags to the build lanes (see Chapter 5).

8.4.2 THE BUILD LANES

The Build Lanes (BL) are buffer lanes that are used to store bags (of each flight/segregation) in front of the build cell (see Figure 8.5). Once a certain number of bags (equal to the ULD capacity) and the ULD are ready, the build process can start. The build process can also start at a defined time (schedule prepared by the automated batch build management system). The purpose of these lanes is to keep the flow at the build cell as continuous as possible. One build lane length is around 42 m = 35 (bags) * 1.2 m/bag.

The number of build lanes per build cell is minimum 2 and it mainly depends on the distance between the baggage storage area and the build area, the build time, etc.

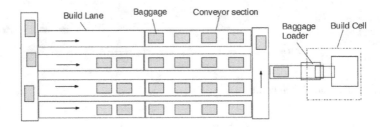

Figure 8.5 Build lanes schematic

8.4.3 THE EMPTY AND FULL UNIT LOAD DEVICES

The ULDs (empty and full) are transported to/from the build area using tug and dollies, roller conveyors, etc. Figure 8.6 shows a general configuration of a build area including the ULDs flows.

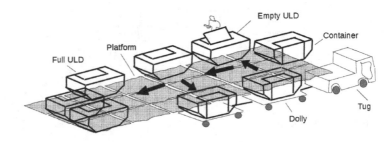

Figure 8.6 Empty and full ULDs transport

The ULDs flow can be summarised as follows:

1. Transport a ULD from empt ULDs buffer into the build area.
2. Load a batch of bags (flight/segregation) into the ULD.
3. Transport the full ULD from build area to the full ULD buffer.
4. Transport the full ULD from the buffer to the aircraft stand.

One of the approaches is to position the ULD in front of the build cell. The ULD is filled-in on it corresponding dolly (see Figure 8.7). The Empty ULD are brought to the area 2 by 2. Tug+2 are positioned in front of the build area. Once the first ULD is full, the dolly is moved forward to allow the operator to load the second ULD. Once the 2 ULDs are full, the Tug+2 is released to transport the ULDs to their destination.

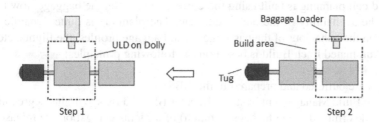

Figure 8.7   Tug+2 in front of build area

8.4.4   THE BUILD CELL STATION

The Build Bell (BC) is foreseen to load bags into a ULD (see Figure 8.8). It is allocated for a certain period of time to a given combination flight/segregation.

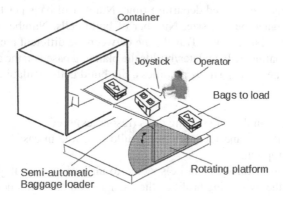

Figure 8.8   Build cell station

The build cell can be: purely manual (i.e., operators loading bags), semi-automatic (i.e., loader, Semi-automatic baggage loader, vacuum, etc.), or fully automatic (i.e., robots, APC, etc.). The batch size is about 35 bags, which is the average number of bags that can be stored in an ULD.

The standard process at a build cell is as follows:

1. The empty ULD is positioned and locked in front of the station.
2. The operator opens the ULD's door and.
3. The operator starts loading the bags into the ULD.
4. Once, the ULD is full, the operator closes the ULD door.
5. The ULD is released and moved out of the build cell area.

8.5   THE BATCH BUILD MANAGEMENT

The build cell planning is built using the current flight data, the baggage flow in the system, the availability of the build cells, etc. The planning is quite dynamic as it depends on the behaviour of the passengers and baggage profiles, the flights, etc.

The Automated Bach Build is based on the following principles:

-   Bags are buffered and prepared in the BBS.
-   Batch Build Management System allocates BLs and BCs to flights/segregations.
-   At a determined time, the bags (a batch) of a flight/segregation are released from the BBS into a BL (A travel time is the time needed to collect all the bags of a given batch).
-   The bags are stored in the BL per flight/segregation (close to the allocated BC).
-   At a determined time, a batch of bags is released from the BL into the BC.
-   The Operator loads the bags into the ULD.

The following constraints are considered to allocate flights/segregations to build cell stations : Flights estimated departure time, Number of bags per flight, Number of operators, Baggage buffers size, Number of build cells, Number of build lanes (minimum 2), ULD buffers size, Travel time between the different areas, etc.

The IT layer continuously updates/builds the planning based on the available bags, the flights schedule, empty ULDs readiness, the build cells availability, etc. with a target of :

-   Minimise the number of building cells (resources) used.
-   React in real time and reorganise the build schedule in case of internal or external disruption.
-   Build strategy (strategies can be specific to airlines or even flights) can be adapted on the fly making profit of the changes and reducing their impacts.

8.5.1   THE EARLY BATCH BUILD

The number of ULDs to build depends on the number of bags, the flights/segregations and the baggage profile (input of check-in and transfer bags into the system). The bags that cannot be loaded into the ULDs using build cells (pull approach) are sent to MUPs (on-time bags), or to the expedite area (late bags).

The early bags are stored in the BSS (reasonable size). The MUPs are open 210/150 min before the STD. The batch build concept is used for a set of flights/segregation from STD-X till STD-Z. Figure 8.9 shows an example of the early build allocation of the flights.

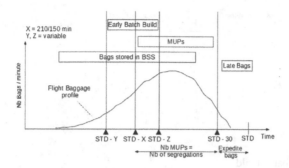

Figure 8.9 Early batch build and MUPs

The X, Y and Z are dynamic variables and depend on the BSS capacity, flight/segregation baggage profile, build cell occupation, number of MUPs available, etc. The X depends on the BSS capacity, number of flights/segregations, baggage profiles, etc.

8.5.2 THE EARLY AND LATE BATCH BUILD

Figure 8.10 shows an example of the early and late build allocation of the flights/segregations. The early bags are stored in BSS (reasonable size), and the MUPs are open 90/60 min before STD. The batch build concept is used for a set of flights/segregation from STD-X till STD-Z.

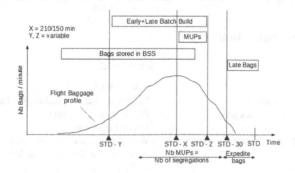

Figure 8.10 Early and late batch build and MUPs

8.5.3 THE ALL BATCH BUILDS

Figure 8.11 shows an example of the all-build allocation of the flights. Early bags are stored in BSS, the batch build concept is used for all flights/segregation, while the too late bags are sent to expedite chutes.

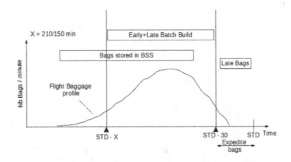

Figure 8.11   All batch build

8.6   THE IMPACT OF THE BATCH BUILD CONCEPT ON THE BHS

Table 8.1 shows a comparison between the conventional and batch-build concepts. Figure 8.12 shows some batch-build general facts:

1. The required BSS capacity needed decreases when the number of build cells increases.
2. The number of operators decreases when the number of build cells increases.
3. The baggage flow increases when the number of build cells increases.
4. The number of bags and ULDs handled increases when the number of build cells increases.

Table 8.1
Conventional vs. batch build

Conventional	Batch build
- High OPEX	- Low OPEX[2]
- Many Make-up or Laterals	- Reduced number of Make-up or Laterals
- Full utilisation only at peak times	- Best utilisation all times
- Huge manpower	- Limited manpower
- Workload dis-balanced among operators	- Workload balanced among operators
- Manual loading	- Automatic loading (or manual/semi-automatic)
- Some miss bags	- No? miss bags

---

[2]But probably higher CAPEX due to the required structure of the BHS installation.

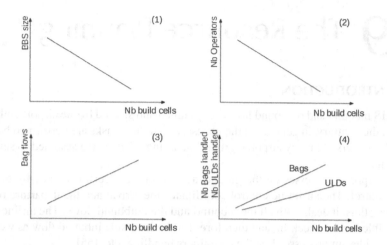

Figure 8.12   Batch build general facts

8.7   CONCLUSIONS

It is known that there are four major players in the field of baggage handling: passengers, airports, airlines and baggage handlers. The passengers, airports and airlines may benefit from the utilisation of the pull concept, whereas baggage handlers may have divided opinions. On one hand, the handlers may support the pull systems since these systems improve the BHS capacity; on the other hand the pull systems will decrease their freedom with respect to planning of the work; thus, they may lose control/power since the system itself decides what and when it will build bags [96].

Changing the BHS from a push system to a pull system, the aircraft will be the instigator by requesting its set of bags and the aim of the BHS will be to deliver all the corresponding bags as efficiently as possible [129]. The benefits of batch building include:

- Speed loading – all bags for one departure are handled at one time.
- Efficiency – while a first batch is loading, a next batch is in preparation, etc.
- Space saving – reduces the needed make up areas.
- Optimising resources – resources can be minimised and optimised.
- Health and safety – reduces physical work (3).
- Operator controlled – the operator determines when to pull the batch.
- Flexibility – the operator can control batch building in different ways.
- etc.

9 The Resource Planning

9.1 INTRODUCTION

The BHS inbound and outbound flows are planned and steered by the airport authorities. On the contrary, in general all the necessary handling tasks are performed by the ground-handlers. At many airports, ground handling services are assigned to several companies.

The airport's processes on the ground depend on the decisions made by the air traffic control. The air traffic control coordinates the arrival and the departure times of planes; thus, it deals with all the inbound and the outbound flows. The airlines are responsible for the check-in, and therefore, for the land-side inbound flow as well as for other relevant processes like the containers handling, etc. [55].

Thus, there are many entities involved in the planning processes (optimisation problems). Each of them managing its own resources, this leads in general to local optimums. The objective is how to deal with all the flows and capacities, etc. in an efficient way. The Resource planning problem covers resources related to both outbound and inbound processes in the airport. The outbound processes include check-in desks, baggage storage, make-up, chutes, etc. The inbound processes include reclaim belts, stands, parking, gates, etc.

9.2 BHS RESOURCES

There are many publications in the field of resource planning for airports. Abdelghany et al. [2] explain the different functions and tactics performed by airlines during their planning and operation. Marín [115] studied the taxi planning at airports. Justesen et al. [89] studied the allocation of ground-handling resources at Copenhagen Airport. Table 9.1 provides an overview on how the different actors that influence the three baggage handling processes (inboud, transfer, and outbound).

By default, staff do not like frequent resource reallocation; they prefer to keep the same configuration for a long period of time (a couple of months or years). It is a human resistance[1], resilience to changes. Scheduling check-in desks, parking, chutes, builds, laterals, reclaim, carousels, storage, etc. are challenging problems.

The assignment of aircraft to stands is an important part of the daily operation of an airport. The location and the type of stand allocated to a plane affects the service performed by several departments within the airport such as catering, customs and passport control, etc. Certain factors such as size of aircraft, origin of the flight (domestic, international) and its destination, the availability of stands, etc. determine the assignment of the resources.

---

[1]Too many frequent changes or requesting too much flexibility from the handlers can lead to strike movements.

DOI: 10.1201/9781003432920-9

Table 9.1

Airport actors and the baggage handling processes

Actors (Tasks)	Outbound	Transfer	Inbound
Airport operators			
Outbound handling	x		
Inbound handling			x
Transfer handling	x	x	x
Flight/Gate parking	x	x	x
Baggage infrastructure	x	x	x
Customs			
Customs		x	x
Border control/immigration		x	x
Airlines			
Check-in	x		
Container sorting	x	x	
Ground handlers			
Container handling	x	x	x
Staffing	x	x	x

It is crucial that the operations are carefully scheduled to prevent malfunctions in the BHS. Prior to determining the BHS operational policies, an important step is the the estimation of system operating capacity. This ensures that the check-in stations are not overloaded with bags, which can cause cascade stoppages and blockages to sub-conveyor sections. Cascading blockages can potentially lead to a poor level of service; customers may depart without their bags, flight delayed, etc.

A set of rules for resource planning exists. Thus, some constraints are hard and must be satisfied, while others are soft and should be met if possible. There is no standard allocation procedure that can apply to all airports. In general, some rules of thumb are used, one seeks only a valid solution and not an optimal one!

The optimal allocation is based on anticipation, as an example from recurrent flights data (of a season). It can be based on a stream-flow, by updating confirmed flights originating from the Flight Information Management System (considering time differences, additions and cancellations). Some of the most known strategies are the usage of all available piers, minimise the pier usage, etc. [40].

9.3 THE RESOURCE PLANNING

The resources planning problem covers resources related to both outbound and inbound processes in airport. The purpose is to allocate resources to flights, given a set of constraints, the optimisation criteria, and a set of objectives. The main constraints are the physical layouts and the locations of the different resources (e.g., check-in desks, ticketing offices, reclaim belts, etc.).

Search algorithms define the problem in terms of a search space filled with a set of points (solutions). The problem is then transformed into the problem of searching for the best solution(s) somewhere in the space of valid ones. In general, the search procedure is composed of three steps, which are (1) describe the problem, (2) define the goal and (3) and use a method to reach this goal. To tackle a search problem over some space of possible solutions, it is necessary to construct a representation (encoding) of the possible solutions for manipulation and storage.

The resource planning problems are known as NP-hard problems and have been studied for several decades without yielding an easy method, but it is widely believed that no such method exists to solve them optimally. As classical methods (e.g., Simplex, Constraints Programming, etc.) are time consuming for larger and non-linear problems, the emphasis is put on the non-deterministic research techniques such Genetic Algorithm, Simulated Annealing, Tabu Search, etc. (see Chapter 14).

The objective is to optimise an economic function, such as, minimisation of distance to the ticket office of the involved airlines; minimisation of the distance to piers, maximisation of the resources utilisation, the revenues optimisation, etc.

9.4   THE PROBLEM ASSESSMENT

The aim is to allocate check-in desks, chutes, build cells, and reclaim belts, stands, etc. to out- and inbound flights (N-allocation problem). Airports must perform the allocations of all the resources (global solution), to optimise their utilisation [55, 71].

It is possible to display the allocation data into a Gantt chart (graphical representation) (see Table 9.2). Via the Gantt chart, the resource planning supervisor can view and edit the allocated resources. A flight can be assigned to different sorts of resources (chute/lateral/carousel, etc.).

Table 9.2
Chute allocation example – Gantt chart

	09:30	10:00	10:30	11:00	11:30	12:00	12:30	13:00	13:30	14:00	14:30	15:00	15:30
chute1					A	A	A	E	E				
chute2					A	A	A	E	E				
chute3						B	B	B					
chute4						B	B	B					
chute5				C	C	C	C	F	F				
chute6				C	C	C	C	F	F				
chute7				C	C	C	C	F	F				
chute8						D	D	D	G	G	G		
chute9						D	D	D	G	G	G		
chute10						D	D	D					
chute11						D	D	D					
chute12													

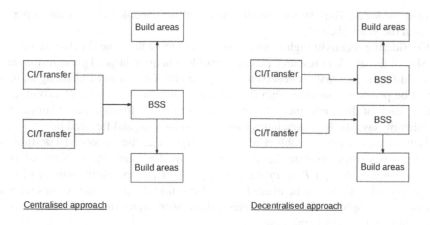

Figure 9.1 Centralised/decentralised baggage storage system

9.4.1 THE REQUIRED SCALABILITY

While the allocation of check-in desks and chutes may appear only loosely related, the two problems are in fact interdependent. Indeed, the topology of the conveyors that move checked-in bags away from the check-in desks may make some chutes inaccessible[2] (at least difficult) for bags checked-in at certain desks. Thus, the way the check-in desks are allocated to flights will influence the way the chutes can be allocated to flights. In extreme cases, the best check-in desk allocation could allow for only mediocre allocation of the chutes, resulting in a poor overall performance of the outbound passenger handling process. A centralised (not shared) baggage storage solution, may lead to poor utilisation of the system, etc. (see Figure 9.1).

Consequently, an adequate resource allocation tool should be designed in a way that it will enable the airport, when the need arises, to optimise the allocation of the check-in desks and baggage chutes simultaneously [75]. With this in mind, and knowing that the number of flights and passengers is bound to increase over the years, the appropriate tool must offer a high degree of scalability.

9.4.2 THE PROBLEM SIZE, STRUCTURE, AND STANDARD SOLVERS

In general, the check-in resources allocation alone involves a hundred resources and a hundred daily flights. Both numbers can be expected to grow over the years. Furthermore, it is reasonable to expect that the airport has at some point to attempt to allocate the resources over longer periods of time than a single day, say a month or a season. We can therefore expect that the allocation algorithm will face at some point the following problem "allocate 100 check-in resources to 200 flights per day, in a

---

[2]This depends on the topology, thus, on the way the BHS installation has been designed.

seven-days' week. Thus, striving to offer an algorithm capable to satisfy the airport's needs in the years ahead!".

Considering that most flights can be assigned to more than one check-in resource, the size of the check-in resource planning problem is quite large. This is further aggravated by the fact that, as observed above, the check-in resource planning should in fact be performed together with (i.e., at the same time) the chute allocation, etc. Indeed, each of them influences the other ones somehow. Ideally, the additional dimension of, say, *100* baggage chutes should be considered and factored in.

With an optimisation problem of such a large size, the important question is whether it has some structure that could be exploited by standard problem solvers, such as *Simplex, Integer Programming*, etc. [157]. Thus, the clear majority of uninteresting allocations will be filtered out and can find the high-quality ones within a reasonable response time. Nevertheless, there is a strong indication that such a structure may not exist in this case.

The Simplex algorithm handles integer variables (i.e., discrete), nevertheless it crucially relies on the assumption that those variables can be "linearised" in some way. For example, the suite (1, 2), (2, 4), (3, 6), etc. can be *"linearised"* as (x, 2*x), and the Simplex algorithm can indeed take a significant advantage of that fact. However, this appears not to be the case in the problem faced by most airports, because the main variables, namely the check-in resources, lack even a concept of a natural order (*let alone linearity*), since renumbering the check-in resources in a different way still yields the same allocation problem.

Let's take flights 1, 2, 3 and resources 2, 4, 6, in which (flight 1, resource 2), (flight 2, resource 4), (flight 3, resource 6), etc. has the same meaning in the present problem as (flight 1, resource 4), (flight 2, resource 6), (flight 3, resource 2), etc. Simply, if the resources are re-numbered as follows: *resource 2* by *resource 4*, *resource 4* by *resource 6* and *resource 6* by *resource 2*. While the suite: *resource 2, desk 4, resource 6* (first resource numbering) may appear to be linear, the suite *resource 4, resource 6, resource 2* (second resource numbering) does not.

Therefore, it can be concluded that this resource planning problem is not easily amenable to *linearisation*, as it is fundamentally *non-linear*. Thus, it can be expected that the Simplex algorithm and other standard solvers may have difficulties solving the problem at hand. In practical terms, such solvers can be expected to spend an inordinate time finding high-quality allocations, yielding response times so long (may be hours) that they would be unusable. It is worth noting that this goes directly against the requirement of scalability identified in section 9.4.1, and may disqualify the linear programming approach from a *"what-if"* analysis scenario.

9.4.3   A NON-LINEAR, YET EXPLOITABLE STRUCTURE

It appears that this resource planning optimisation problem lacks a linear structure that could be exploited by standard linear programming solvers. This does not mean that the problem lacks any exploitable structure – it just means that it lacks a structure necessary for a successful application of standard linear programming solvers.

In fact, the problem does have a structure that can be exploited by an adequate optimisation algorithm. Indeed, let us consider the objective of the allocation: check-in resources (chutes, and reclaim belts) must be allocated to flights, under several constraints, in an optimal way. In other words, flights must be grouped together, into groups that can be handled by a check-in resource (a chute or a reclaim belt). The problem consists of grouping flights together, and this grouping aspect of the allocation is a structure that can be exploited. This is illustrated in Figure 9.2, where the check-in resource planning on the left is equivalent to the grouping on the right.

Therefore, it is possible to design an algorithm exploiting that structure, delivering high-quality allocations within short response times. It is, however, well known, that the linear programming approach used in the Simplex and other standard solvers, is unable to take advantage of this structure because it is a grouping structure, and not a linear one. Simply put, linear programming solvers are not designed to take advantage of the grouping structure, and a different technique must therefore be used[3].

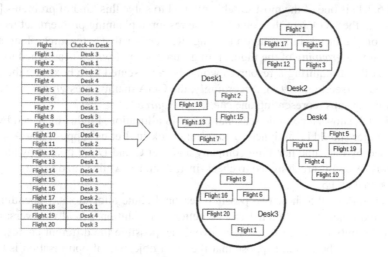

Figure 9.2   A check-in resource planning and the corresponding flight grouping

9.5   THE GROUPING PROBLEMS

As observed above, the allocation problems faced by airports exhibit a grouping structure. Flights are to be grouped together into groups assigned to (a set of) check-in resource(s). Flights are to be grouped together into groups assigned to a chute. Flights are to be grouped together into groups assigned to a reclaim belt, etc. In

---

[3]The observation regarding linear programming is not meant to put in doubt the performance of the Simplex and other standard solvers – they are excellent techniques. But they need to be applied to problems with a structure they can exploit, otherwise, they will not perform adequately.

Table 9.3
Seasonal check-in allocation plan

	08:00	08:30	09:00	09:30	10:00	10:30	11:00	11:30
check-in 1		A222						A5555
check-in 2		A222						
check-in 3				B234				
check-in 4				CB35		D3333		
check-in 5		D1234						

fact, all these problems belong to a class of optimisation problems called grouping problems (see Chapter 14). The Grouping Genetic Algorithm (GGA) (see section 14.6.7.1) is one of the most suitable method to solve this kind of problems [52].

To place the GGA into the context of the resource planning problem, let us concentrate on the check-in resource planning. Remember that a resource planning is equivalent to a *grouping* of the flights into groups, each of which is assigned to a resource (A flight requiring *n* resources would be represented by n boxes on the right of the figure) (see Figure 9.2). Accordingly, the GGA manipulates *groups of flights*, each of the groups representing one *check-in resource*.

Table 9.3, shows a possible seasonal check-in allocation. Flights managed by the same handler should be assigned to adjacent check-in. Let us suppose that A and B are treated by the same handling company and that C and D are treated by another one. Thus, A and B are assigned to check-in 1, 2, and 3, while C and D are assigned to check-in 4 and 5.

Tables 9.4 and 9.5 show two possible seasonal chute allocations. Depending on the selected objective, solution 1 can be better than solution 2 (and vice-versa). Indeed, if the only objective is to use as evenly as possible the different chutes, then solution 1 is the best. But, suppose that the main objective of optimisation is to reduce the number of the used chutes, then solution 2 is the best, etc.

Table 9.4
Seasonal chute allocation (possible solution 1)

	08:00	08:30	09:00	09:30	10:00	10:30	11:00	11:30
chute 1		A222						A5555
chute 2								
chute 3				B234				
chute 4				CB35		D3333		
chute 5		D1234						

Table 9.5

Seasonal chute allocation (possible solution 2)

	08:00	08:30	09:00	09:30	10:00	10:30	11:00	11:30
chute 1		D1234		CB35		D3333		
chute 2		A222		B234				A5555
chute 3								
chute 4								
chute 5								

In the process of resource planning, we encounter several constraints that arise from various aspects of the passengers and baggage-handling process. Some examples of these constraints include:

1. *capacity constraints*: each resource has a finite capacity to accept flights. Example: no resource can be open for longer than 24 hours a day.
2. *associative constraints*: if some flights are assigned to a set of resources, then some other flight(s) must also be assigned to that same set of resources.
3. *dissociative constraints*: if some flight are assigned to a set of resources, then some other flight(s) cannot be assigned to that same set of resources.
4. *contingency constraints*: when a flight with a certain property (handler, zone, etc.) is assigned to a resource, then only flights having that same property can also be assigned to that resource.

9.6   CONCLUSIONS

The system dynamic routing can change the BHS daily operations (routing and allocation) from stressful reactive tasks into a proactive monitoring function, where the operator is warned of any condition in the system that requires action before the performance of the system process is affected. Based on the provided information by the system, operators are enabled to adjust the operational tasks to prevent undesired situations arising, and even to improve the overall performance of the system.

There is no doubt that the resource planning optimisation problem faced by the airports features several different constraints that a valid allocation must comply with. It has been shown that the GGA algorithms can cope with a very wide range of possible constraints. Importantly, the compliance with the problem's constraints is all that is needed for the GGA to deliver an optimising power superior to other techniques, since the resource planning problems faced by the airports are in fact *grouping* problems.

10 The Service Vehicle Management

10.1 INTRODUCTION

More and more, airport IT systems, air traffic controllers, and airlines, etc. tend to communicate seamlessly with each other. Often, this inability to share real-time information causes flight delays, long turnarounds, and idle run-way time, etc. New technologies can facilitate collaborative decision-making to optimise traffic and runway management, reducing runway wait times, efficient baggage sorting, baggage storage, etc.

The aircrafts that land at an airport are parked at a gate or stand and are served by a set of service vehicles, as shown in Figure 10.1. An airport service provider is used in meeting demands for services from arriving and departing aircraft. The quantities of vehicles and employees are determined by the requested quality. Mastering how to meet a set of demands is a management and a scheduling problem (see Chapter 14).

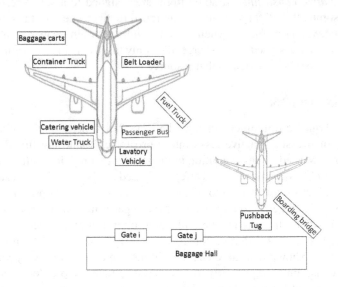

Figure 10.1 Resources involved in service delivery

The stochastic demand (i.e., number of passengers, number of bags, containers, aircraft, etc.) is one of the major challenges for the airlines, airports and handlers in their quest of continuous improvements. One of the problems the handlers face

DOI: 10.1201/9781003432920-10

is the Service Vehicles Management (SVM). Many dispatching and optimisation techniques have been developed to manage the vehicle movements [48, 64].

In big airports, the service vehicles typically almost do not return almost to a central depot (or any other common location), before the end of the working shift. Service providers must plan routes that involve spending varying amounts of time at requests' locations (e.g., parked aircraft, docks, buffers, etc.) and in general, unfortunately they cannot serve multiple requests at once [101].

This problem involves managing a fleet of service vehicles with a complete knowledge of which aircraft should be serviced. Nevertheless, an incomplete knowledge is provided of when and where these aircrafts will be requesting a service and how long it will take to service each of them [101]. The system's available data (e.g., requests, resources, etc.) is updated over time, so that a dispatcher (operator or an IT system) knows more-or-less the demands for the services for the next x minutes (let's say 30) but not so precisely those of the next y hours (let's say 3).

10.2   THE MATHEMATICAL FORMULATION

The service to an aircraft is formulated as a set of requests (Figure 10.2):

1. Outbound: from aircraft to a set of other locations, and
2. Inbound: from a set of locations to the aircraft.

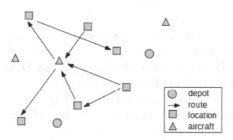

Figure 10.2   Aircraft and vehicles services

Mathematically, the problem can be defined as follows. Let N be a set of services/transportation requests, and for each request $(i \in N)$, a load (of type t_i) and of size q_i should be transported from an origin O_i within the time window $[Ostart_i, Oend_i]$ to a destination D_i within the time window $[Dstart_i, Dend_i]$ as illustrated in Figure 10.3.

The requests are subject to precedence constraints (i.e., some requests cannot be handled before others). Each customer[1] c has a location $(x_c; y_c)$. The V vehicles will serve these N requests. Each vehicle $(v \in V)$ carries k products (items or services)

---

[1] In this context, the customer refers to the aircraft or flight, which requires a range of services such as parking, catering, baggage handling, and more.

and travels a total distance d_v: from the depot, to all its q customers (the set $\{c_{v1} \ldots c_{vq}\}$) and back to the depot (end of the working day, end of shift, break, etc.). Each customer receives his amount of services q_i. A solution is feasible only if all customers receive their service within their time windows.

Figure 10.3 Time windows constraints – example

Amongst others, the following data is needed:

1. Fleet: a set of tugs, trucks, etc.
2. Map (routes): for each pair (i, j) of locations, the estimated travel time (t_{ij}) and the corresponding distance (d_{ij}) should be measured on the network.
3. Missions: a set of requests generated by the different aircrafts.

A route R_k for a vehicle k can be regarded as a directed route through a subset $S_k \subset S$ such that:

1. R_k starts and ends at the corresponding depot (each depot covers its surrounding area) (Figure 10.4);
2. Each O_i location is visited before the corresponding D_i location;
3. Vehicle k can visit some locations more than once;
4. The vehicle k load's never exceeds its capacity Q_k;
5. A set of time intervals called vehicle availability (vehicle j is available from (begin time)$_j$ to (end time)$_j$).

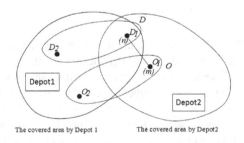

Figure 10.4 Graphical representation of the problem to tackle

The main outputs are:

1. For each vehicle, there is a route manifest which contains a detailed list of stops in a chronological order.
2. For each stop, there is a list which gives; the starting location, the process, and the time allocated for the vehicle at this stop (estimation).
3. The times of departure and arrival to the depot.
4. Etc.

Among many others, the following indicators can be analysed:

1. Flow between the different locations
2. Routes utilisation
3. Areas utilisation optimisation
4. Resources inefficiencies
5. Resources utilisation balancing
6. Storage zones capacity
7. Traffic between the different zones
8. Number of vehicles used
9. Quality of service
10. Delays (if any!)
11. Number of operators allocated to the different tasks
12. Etc.

10.3   THE MANAGEMENT AND THE RULES

A problem involving the management of a fleet of vehicles is considered here. Several aircraft are requesting services (e.g., water, catering, fuel, baggage, passengers, etc.) at various times and locations (e.g., depots, BHS facilities, stands, ULD Storage, etc.) over the course of a day. The objective is to schedule a fleet of vehicles so that each aircraft is serviced on time [166]. Thus, firstly, determining the initial routes, and secondly, inserting the non-served trips (i.e., trips that have not been carried out due to the time constraints).

In general, the dispatcher must monitor regularly the updated data regarding the arriving and the departing aircrafts. When an aircraft is (let's say 15 minutes), away from requesting services, the dispatcher orders the vehicle that has been idle the longest (if any) to go immediately to a position where it can provide service to the aircraft. If there is no idle vehicles, the dispatcher waits until the first vehicle becomes idle and then assigns it to this task, etc. [101].

The task of assigning service requests to vehicles is practically difficult and can be considered as a complex Combinatorial Optimisation Problem (COP) as the number of routes increases as a function of vehicles and requests. For example, requests with geographically close destinations are likely to be carried out by the same vehicle, but some destinations may have origins that are geographically far apart and it is impossible to carry out the transportation requests by the same vehicle. Therefore, the best solution for achieving an ideal transportation request is based on the minimum number of vehicles while satisfying all the requests.

Figure 10.5 presents a vehicle routes solution, in which the circle represents the depot from where the vehicles depart. The squares represent the various customers (work location) and the connecting lines (arrows) represent the vehicle routes. To reduce visual clutter, there are no lines leading to and from the depot.

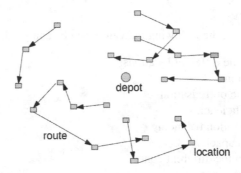

Figure 10.5   Vehicles routes

The offline approach deals with the static problem in which all the requests are known at the time of constructing the routes, while in the dynamic one (real time), some requests become available in real time during the execution of the tasks (see Figure 10.6). The simplest form of the dynamic problem is often solved as sequence of static problems. In this form, each time the current routes are updated whenever a new request of service becomes available.

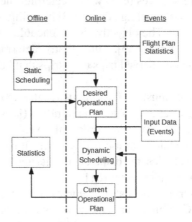

Figure 10.6   Offline and online scheduling

10.4   THE METHODOLOGY

For each i (request) $\in M$ (number of used vehicles), the time window $[e_i, l_i]$ denotes the time interval for carrying out the provided service at a given location. Given the service plan (i.e., pickup and delivery) and departure time of the vehicles, the time windows defines for each request i the arrival time A_i and the departure time D_i [167].

As illustrated in Figure 10.7, if $A_i < e_i$ (i.e., the vehicle arrives before the time determined by the customer, A_{i1}), the vehicle should wait at that location till $A_i = e_i$. If $e_i < A_i < l_i$, (i.e., the vehicle arrives to its destination within the time allocated by the customer, A_{i2}), then the customer is served directly without any waiting time and this is the best situation. If $A_i > l_i$, (i.e., the vehicle arrives after the time determined by the customer, A_{i3}), this vehicle cannot be used, and this represents the worst case as a new vehicle should (already) been ordered to satisfy that request.

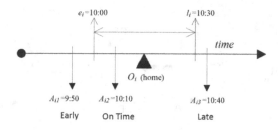

Figure 10.7   The time windows for a service plan

After initialising the available number of vehicles, each route is initialised with the earliest customer (i.e., the customer who has the earliest time at the origin location). The cost to assign a new customer to a vehicle depends on its position in a route. In other words, if the customer is the first to be served by the vehicle, then this cost represents the travel time from the depot to the location of this customer. This insertion cost of customers in routes permits grouping customers with the same origin and destination.

The non-served requests are iteratively inserted into the current routes. A new route is initialised when non-served requests cannot be feasibly inserted into any of the N currently active routes. Let n be the current number of non-served customers. The assignment of the non-served customers to the current routes is obtained by using the "Best Fit" approach. If a customer cannot be served in a route (e.g., time window violation), its cost will be set to infinity (to prevent its insertion in this route). If a customer cannot be feasibly inserted into any of the current routes, then a new route is started by selecting the best-unused vehicle that can carry out the trip. This selection is based on the vehicle efficiency score and the average distance from the vehicle depot to the customer location. In the case of a non-served customer that cannot be feasibly served either in one of the current routes or in a new one, then at least the following two situations can be encountered:

1. The number of vehicles is fixed (this customer cannot be served by this solution). Thus, the aim is to minimise the number of non-served customers.
2. The number of vehicles is not fixed (or there is a possibility for hiring vehicles). Thus, the aim is to minimise the number of used vehicles.

10.5 THE LOCAL IMPROVEMENT HEURISTIC

The local improvement procedure (Farthest heuristic) can be coupled with some algorithms to accelerate the search process for obtaining an ideal service plan. Thus, the aim is to provide the served customers with better quality service by minimising the ride time. Having the same destination and nearest origin for a given set of customers, the aim is to serve the customer who has a far location (from destination) rather than to begin with D1, which will have a long ride time as shown in Figure 10.8.

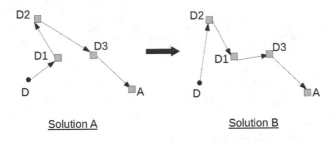

Solution A   Solution B

Figure 10.8 The farthest heuristic procedure

In this case, the vehicle must begin its route with the customer that has the far location, and then to pass to the other locations to decrease the distance from their destination. In this case, the Farthest Heuristic (FH) must consider the time windows and the destination of each of the given customers assigned to that used vehicle. For each 4 successive locations (D, D_1, D_2, D_3), two solutions are possible (see Table 10.1).

IF $(Dist_{A1} + Dist_{A2} + Dist_{A3}) < (Dist_{B1} + Dist_{B2} + Dist_{B3})$
THEN, the best solution is solution X.
ELSE, the best solution is solution Y.

Table 10.1
Comparison of the two solutions

Solution X	Solution Y
$Dist_{A1} = DD_1$	$Dist_{B1} = DD_2$
$Dist_{A2} = D_1D_2$	$Dist_{B2} = D_2D_1$
$Dist_{A3} = D_2D_3$	$Dist_{B3} = D_1D_3$

10.6 CONCLUSIONS

The problem addressed in this chapter is the problem of assigning resources, and the routing of the different vehicles. A solution for the problem is a set of routes $R = \{R_k | k \in V\}$ and the aim is to minimise the objective function $f(R)$ in terms of service quality and the number of used vehicles. This service quality is considered to satisfy the needs of the customers especially within the required time slot. The problem can be easily formulated as a grouping problem since the aim is to combine several requests in a set of routes (see Chapter 14).

Thus, the vehicle's load is the sum of the q_i which corresponds to the requests that are served by a given vehicle. The objective function is the number of vehicles V. If V is equivalent for two solutions, then the aggregate distance that the vehicles travel is used as a tie-breaker. The main goal is to find a solution with minimal cost.

The simulation procedure can be used to analyse various scenarios, and to mitigate the risk associated with the dynamically changing environments (e.g., weather, strikes, routing, new airlines, baggage flow, flights delays, constraints, etc.). It can be used to evaluate, plan or redesign systems, new dispatching rules, identify inefficiencies in existing processes and test a variety of scenarios.

11 The BHS Disaster Management

Disasters can result from any catastrophic events (such as earthquakes, floods, volcanoes, pollution, pandemics, etc.) and can affect any region of the world. The lack of familiarity with the characteristics of disasters and their causes is one of the things that exacerbate their effects, widen their reach and their destructive dimensions [150]. Figure 11.1 shows some of the causes of disasters.

Figure 11.1 Causes of disasters

Disasters have negative social, economic, etc. impacts on individuals and societies. The losses (humans and materials) resulting from these disasters has increased in the last 50 years. Therefore, many research institutions specialised in the field of disaster management try to deal with the risk of disasters and their impacts. They use many empirical models, scientific tools, several technologies to process and analyse data on these disasters [109].

Airports also, must be ready for the unexpected events, with dedicated airport emergency plans and regular drills that are necessary for building their capacity to cope with disasters and crisis. Polater [140] presented the literature base of airport disaster management for non-aviation-related events. He reviewed the literature to

DOI: 10.1201/9781003432920-11

Table 11.1
Issues and potential solutions

Issues	Potential solutions
Flooding	Topology of the land and position of the building
Fire	Equipment redundancy
Terrorism	Fencing around the building
Pandemics	Various solutions
Accidents	Training, H&S
Cyber-attacks	Security
Sabotage	Security
Drones attacks	Security, fencing
Bags die-backs	Maintenance, re-routing capabilities
Complex projects	Divide projects in steps
Network	Redundancy
Hardware	Redundancy, Quality
Software	Quality checks, Upgrades, Redundancy mainly via distributed redundant systems
IT viruses	An-viruses
New Technologies	Intensive testing before implementation
Systems Failures	Maintenance
Data loss	Backups (+cloud)
Personnel	Training, Incentive
Ergonomics	Manual Handling Aids
Capacities	Mix of operators and robots

report Airport Disaster Management (ADM) efforts, identify the existing gaps and to determine the main questions to be addressed by the research community.

A flood, earthquake, etc. can affect the whole airport. On the other hand, the loss of a sorter or any other resources of the BHS could be more judiciously approached: provide for each process a fallback solution, may be a manual one! Thus, airports have to be ready for the unexpected, with dedicated airport emergency plans and regular drills essential to their ability to cope with should disaster strike. Table 11.1 gives elements of solutions to some issues that can be raised following a disaster.

The impact of the recent pandemics on air travel has been dramatic, making it the worst aviation crisis ever. The perspectives for recovery of air travel are bleak, with an estimated return to the normal traffic level to take at least a couple of years.

The topic of airport disaster management deserves more deep coverage than these few lines, as it is beyond the scope of this book. It warrants a dedicated book to cover the subject in detail. I would like to draw the baggage handling business's attention to this matter.

Part III

Tools

12 The Lean Principles

12.1 INTRODUCTION

The concept of continuous improvements is defined as a continuous effort to improve products, services, processes, etc. These efforts can be small step-by-step "incremental" improvements over long time or fast "breakthrough" improvement all at once. Among the most widely used tools for continuous improvement is the four-step model *plan-do-check-act* (PDCA) cycle as shown in Figure 12.1. On the other side, Business Process Management which is a management approach, that helps promoting business effectiveness and efficiency while looking for innovation, flexibility, and integration with other technologies [32, 186, 187].

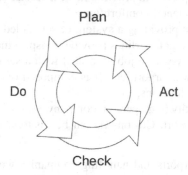

Figure 12.1 Plan-Do-Check-Act cycle

Lean means reduce waste and variability in the process outcomes. In very competitive and changing environments, developing competitive solutions are a huge challenge for organisations. Industry's performance has become multi-criteria matter, and can be economical, environmental, societal, etc. Within this context, organisations that plan to create/update their systems should think ahead about all these objectives to be fulfilled. Therefore, designing and developing a system should be a multi-criteria systemic approach.

The systemic thinking is the process of predicting how something influences another. It is defined as an approach to problem solving, by viewing "problems" as parts of an overall system, rather than reacting to present outcomes or events. The systemic thinking is a framework based on the belief that the sub-systems of a given system can best be understood in the context of their relationships with each other and with other systems, rather than in isolation (view systems in a holistic manner).

The analytical approach is interested in *"of what is done?"*. It works with the microscope to cut off a general system into less complex parties, whereas, the systemic approach replies to the question *"How this is done?"*. It integrates the notion of complexity and focuses more on the interactions between the elements of the system rather than to what constitutes the system.

Solving baggage handling problems (e.g., storage, sorting loading, unloading, etc.) should be seen as one unique problem. Rather than individual separated systems, one should use the analytical and the systemic approaches to find solutions. In the next sections, the main lean principles will be presented, and their applications to the airport businesses will be discussed.

12.2   BHS LOGICAL CONTEXT

From the baggage handling point of view, the logical context of the system is depicted in the following context diagram (see Figure 12.2). The context consists of systems and users, as described below:

1. Airport: An airport is composed of a variety of facilities and infrastructures, where the main purpose is to provide the best service in meeting travellers' needs of safety, efficiency, comfort, etc.
2. Airlines: A business providing a system of scheduled air transport of passengers, bags and freight. Airlines provide transport means, planning, etc.
3. Services: A company whose job is to deal with a set of services provided to customers, airlines, airport, etc. It is composed of resources, operators, transport means, technicians, etc.
4. Customers: Any entity (e.g., person, company, etc.) which asks for services provided by the airport. Customers can be visitors, passengers, baggage items, freight, etc.

In general, airlines, airports and handling companies work together to enhance customers (e.g., companies, passengers, their bags, etc.) processing facilities and to find solutions to improve the communication and the information sharing amongst all parties, including passengers.

Operations and maintenance typically include the day-to-day activities that are necessary for an installation (systems and equipment) to perform their intended

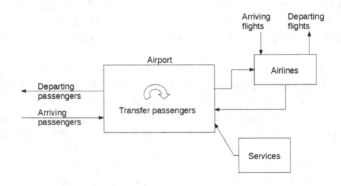

Figure 12.2   Baggage handling systems – Context

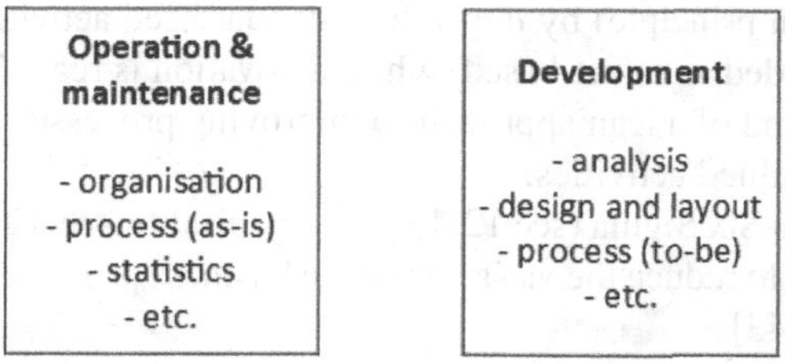

Operation & maintenance	Development
- organisation - process (as-is) - statistics - etc.	- analysis - design and layout - process (to-be) - etc.

Figure 12.3 Operation & maintenance and development activities

functions. The development refers to creative work undertaken on a systematic basis to increase knowledge (e.g., organisational, technology, processes, etc.) and the use of this knowledge to propose new and innovative solutions (see Figure 12.3).

The operation and maintenance deal with the "as-is" processes, the development deals with the "to-be" situation(s). Operation and maintenance handle the resources and operators (organisation), processes (current situation), the current rules, statistics, KPIs, etc. The development is in charge of processes (to-be) and future organisation, design and layout, technical solutions, data analysis and simulation, etc.

To make use of the competences of operation and development teams, a continuous improvement process is the best approach (see section 12.3). An interface team (with some knowledge in the different fields) is needed to facilitate the communication between the different teams.

12.3 CONTINUOUS IMPROVEMENTS

Continuous Improvement (CI) means small improvements, and this is due to the coordinated continuous efforts (see Figure 12.4). It refers to modifications and

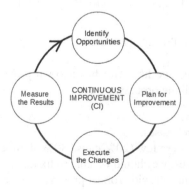

Figure 12.4 Continuous Improvement

continuity and makes better use of the existing resources. The main purpose is to eliminate waste (lean principle) by improving standardised activities and processes. Thus, the CI is regarded as effort-based, while innovation is regarded as investment-based. The CI is a kind of a lean approach to improving processes whereas eliminating waste and non-valued activities.

The CI is based on Six Sigma (see 12.4), and organisations that adopt its principles as a philosophy seek to reduce the variation in the business processes that cause waste and inefficiencies [143].

12.4   SIX SIGMA

Six Sigma (6σ) simply means a measure of quality that strives for a kind of perfection [66, 143]. It is a data-driven methodology for eliminating defects and failures, and it drives toward six standard deviations between the mean value of a process and its nearest specification limits – from product to service.

The statistical representation of 6σ describes quantitatively how a process is performing (as shown in Figure 12.5). A 6σ defect is defined as anything outside of the customer specifications (the boundaries). The 6σ seeks to improve the quality of process outputs by identifying and fixing the causes of defects (errors) and variability in the business processes [84].

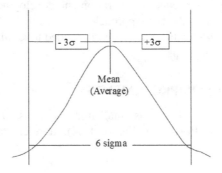

Figure 12.5 6σ – Process statistical representation

As in any industry, baggage handling has its own processes, organisations and technologies. In general all the processes have :

1. A certain Variability,
2. All variabilities have causes,
3. Generally, the causes are few (20% cause 80% of effects),
4. If we know the causes, we should be able to fix them,
5. Then, the designs must provide robust process to the remaining variations,

The purpose of the 6σ methodology is to implement a measurement strategy in order to improve the main processes and to reduce the variations of the outcomes through the application of improvement projects. This is accomplished through the use of two 6σ sub-methodologies: DMAIC and DMADV (see Figure 12.6):

- The 6σ DMAIC process (i.e., define, measure, analyse, improve, control) to improve the existing processes in an incremental improvements manner.
- The 6σ DMADV process (i.e., define, measure, analyse, design, and verify) is used to develop new processes and/or products at 6σ quality levels.

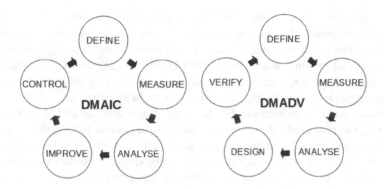

Figure 12.6 DMAIC and DMADV processes

Many research studies and implementations of the 6σ methodology have been implemented in airports. Xinhui [195] presented an application of the 6σ –DMAIC model, to improve check-in service performance in Hang Zhou Xiao Shan airport. Alsyouf [5] presented an implementation of lean 6σ methodology to continuously improve the baggage flow in a BHS, by identifying the causes of the mishandled baggage and deriving solutions to enhance the BHS performance. The applied method illustrate how to avoid baggage congestion and provides applicable and cost-effective solutions.

12.5 LEAN PRINCIPLES

Lean is a practical and engaging way of breaking overwhelming tasks into manageable ones and delivering the improvements [78]. It identifies and eliminates waste to deliver improved values and services, based on identified customer requirements. Thus, it improves existing processes and creates new processes (if any).

The ten rules of lean can be summarised as Eliminate waste, Minimise inventory, Maximise flow, Pull production from customer demand, Meet customer requirements, Do it right the first time, Empower workers, Design for rapid changeover, Partner with suppliers, Culture of continuous improvements [78].

The losses and wastes are potential profits killers, and the goal is to annihilate them. Generally, waste is defined as [194]:

- Any activity which absorbs resources but creates no values,
- Any mistakes which require rectifications,
- Processing steps which are not actually needed,
- Movement without purpose of people and goods from one place to another.
- Resources in a downstream activity standing around, waiting because an up-stream activity has not delivered goods on time,
- Goods and services which do not meet the customer's needs,
- etc.

The application of lean thinking/management aligns and integrates long-term business improvement strategies with day-to-day continuous improvements targets. This ensures that enterprises remain, flexible and ready for the challenges ahead with lowest costs and minimal waste. In the BHS business examples of waste: the double handling of bags, resources (chutes, operators, etc.) waiting for bags, delays, etc. Many industries have adopted the lean principles including airports.

Whenever one tries to increase the efficiency of a system, he will always face an increase of variation. Indeed, new processes leads in general to variation in results due to many reasons such training, resistance to changes, etc. Combining the lean management with the 6σ provides a powerful methodology to drive operational improvements. Its main focus is on improving and eliminating as many steps in the process as possible, and then by reducing the variation in the remaining processes to improve the quality of delivery of the product and services. Applying the lean 6σ can improve any airport planning by making it more visible and predictable.

The chart in Figure 12.7 illustrates the Lean Six Sigma approach, with the X-axis representing process efficiency and the Y-axis representing process performance variation. There are two ellipses on the chart, the first one representing the Six Sigma zone, and the second one representing the Lean approach zone. These two areas are separated by a line. Thus, the lean approach is a methodology used to minimise waste and increase efficiency in a process, while the 6σ methodology focuses on reducing

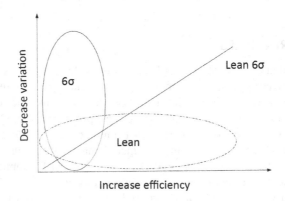

Figure 12.7   Lean and 6σ

variation in process performance. Together, these approaches can help organisations achieve high levels of quality and efficiency.

12.6 BUSINESS PROCESS MANAGEMENT

The Business Process Management (BPM) is an approach that aims to analyse and streamline the processes implemented by an enterprise to carry out its activities with a view to improving its performance. It is concerned with the definition, execution and management of business processes that are defined independently of any single application [186, 187]. Thus, the BPM is a set of work-flows, and further differentiated by the ability to coordinate the activities across multiple applications with fine control [189].

These processes are structured with the daily activities that are necessary for the exercise of any business. They consist of series of tasks and actions (simple or complex), that the employees accomplish in order to obtain a specific result (e.g., check-in bags, security checks, loading bags into ULDs, transporting the ULDs to the aircraft, etc.). The classical workflow is concerned with the application of specific sequencing of activities, involving both automated procedures (software-based) and manual activities. The BPM activities can be grouped into five categories: design, modelling, execution, monitoring, and optimisation (as shown in Figure 12.8) [47].

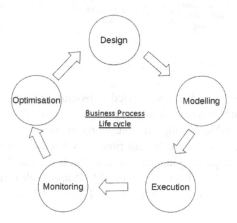

Figure 12.8 BPM steps

The BPM is used to represent the processes of an organisation. It highlights the different interactions and data exchange with the existing information systems. Therefore, it has two major roles: first, to define the enterprise strategy, and second, it is a powerful management approach. The BPM creates value for the business by improving productivity, staff efficiency, and quality of service to the customer.

12.6.1 PROCESS DESIGN

The Process design deals with the identification of the existing processes, "as-is" and the design of "to-be". The focus is on the representation of the process flow, the actors within it, the organisation, the technology, the departments, etc.

The "as-is" is an expression that means the "current figure/situation", and indicates the status to be improved (yes/no). Whereas the expression "to-be" means the targeted ideal situation (i.e., goal).

Improving activities are defined as activities to get closer to "to-be" from "as-is". It is called "ideal figure" or "what it ought to be". Figure 12.9 illustrates an example of the "as-is" and the "to-be" versions of a baggage handling processes.

A good design reduces the number of issues over the lifetime of the process. The aim here is to ensure that a correct and efficient design is prepared. The improvements could be either in human-to-human, human-to-system, and/or system-to-system work-flows.

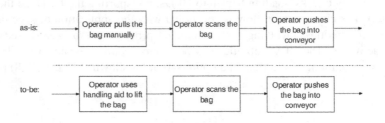

Figure 12.9 Example of an "as-is" and "to-be" processes

Business processes should not be tackled in isolation of any of the other components that make up the dynamic of any organisation. One should not think only of the IT infrastructure and technology that supports the business. One should think of a model that illustrates how the different processes interact with the business components including strategy, people, processes, technology; and the other components that constitute a system. This is known as the T.O.P. approach which is defined as the trio Technology, Organisation and Process (see Figure 12.10).

The main elements are:

1. **Technology**: helps performing the processes. A clear understanding of how the changes (new technologies) would affect the processes and the people.
2. **Organisation** (People): must be a part of any transformation effort. An understanding of both the current culture and the desired one should form a central part of the effort. By describing the desired states in terms of behaviours, one can articulate what processes, technologies, and other capabilities of the organisation one must enable.
3. **Process**: is very much about managing the organisation as a system of interrelated activities to positively influence the outcomes of that system.

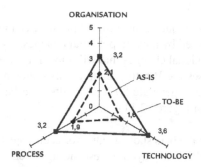

Figure 12.10   From "as-is" to "to-be" considering the T.O.P. approach

In general, the objective is to improve the existing processes and introduce new ones. This includes changes to existing ones, introduction of new ways of working, performance levels and outputs, new products, new teams, new technologies, etc. It covers everything to do with the staff from the organisational culture view. Examples are staffing levels, training needs, culture changes, new skill requirements, etc. It encompasses also systems, tools and other resources such as buildings.

12.6.2   MODELLING

Modelling (i.e., analysis and simulation) is used to implement a theoretical model (a certain design) and introduces parameters that can be used to study the different scenarios (e.g., changes the sequence, changes to the input data, how the system should operate under different conditions, etc.). It also involves running "what-if analysis" on the processes[1].

Computer simulation is an important tool in many areas, especially in cases where the traditional mathematical methods fail in view of the analysis of highly complex systems. In many cases, there are no other methods of investigation at hand, because most analytical methods break down due to the non-linear behaviour of industrial systems, or the experimental methods are too risky, too expensive or infeasible. The simulation topic is introduced in Chapter 15, and it is used to check the results of the design phase.

When many changes are required to move from the "as-is" to the "to-be" situation at the flow-model level, it is difficult for the participants, to keep track of the current flow and the different changes. Thus, work-flow/process engines[2] can be used.

Here after a non-exhaustive list of modelling techniques and tools.

---

[1]Examples of what if?: "What if I have 70% of resources to execute the tasks?"; "What if I want to do the same process for 85% of the current cost?"; etc.

[2]These tools (such as VISIO, ARIS, Draw, etc.) can be used to create diagrams. They offer a wide variety of built-in shapes, objects, and stencils to work with. The driving idea behind is to make visualisation of the process to make it as easy as possible for the user.

12.6.2.1   Pareto Charts

The Pareto analysis is a technique for prioritising possible changes by identifying the problems that can be resolved by making potential changes. By using this approach, one can prioritise the individual changes that will most improve the situation [134].

The Pareto analysis uses the Pareto principle –also known as the "80/20 Rule" –the idea that 20% of causes generate 80% of results. The Pareto diagram consists of two graphs (as shown in Figure 12.11):

- The bar chart, where the bars show a value for each object (number of times an event occurs), and are displayed in descending order.
- The line chart shows the accumulated number of events.

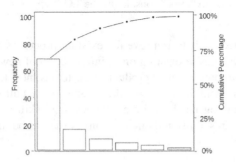

Figure 12.11   Pareto chart

In general, it is difficult to solve all the problems at once, rather start by the most critical ones, the ABC classification method can help on that.

12.6.2.2   ABC Classification

The ABC classification is a ranking system for identifying and grouping elements having the same features. Or, in other words, do these elements grouped together can be useful for achieving some business goals. It is associated with inventory control in warehouses and is used to rank products such as which customers are the most important, which business segments cause the most financial incomes or risks, which employees are the most valuable or which processes are likely to cause issues, etc.

The ABC classification is associated with the 80/20 rule, a metric which suggests that 80% of the outputs are generated by 20% of the inputs (see Figure 12.12). Thus, the goal of this classification is to identify these valuable 20% inputs in order to be controlled most closely.

The system groups things into three categories, (A) extremely important, (B) moderately important and (C) relatively unimportant. Once the A's, B's and C's have been identified, each category can be handled in a different way, with more attention being given to group A, less to B, and even less to C [174].

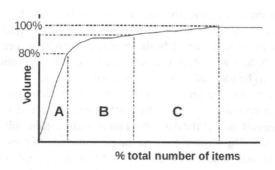

Figure 12.12 ABC classification

12.6.2.3 Fishbone Charts

The Ishikawa diagrams (i.e., fish-bone or cause-and-effect diagrams) are diagrams that show the causes of certain events (see Figure 12.13). These causes are grouped into different categories to identify these sources of variation. The fish-bone diagram identifies many possible causes for an effect or a problem, and it immediately sorts ideas into different useful categories.

A fish-bone diagram is useful in brainstorming sessions to focus on conversation. Once all the possible causes for a problem are identified, they are ranked based on their level of importance (hierarchy). The shape of the diagram looks like a skeleton of a fish. These diagrams are filled from right to left, where large "bones" of the fish are branching out to include smaller bones thus giving more details. The resulting diagram illustrates the main causes and sub-causes leading to an effect (symptom). The systems are then modified in order to solve the issues pointed out by the method.

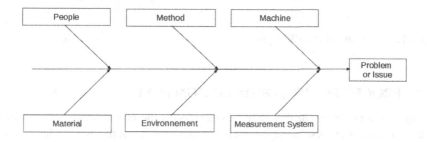

Figure 12.13 Ishikawa diagram

12.6.3 EXECUTION, MONITORING, AND OPTIMISATION

One way to automate processes is to use an application that performs the needed steps of these processes. In general, it is difficult to find applications that execute all the steps of the process efficiently and/or completely[3]. Another approach is to use a combination of human and software which have to collaborate to accomplish tasks. Software for the automation of business process can be directly executed by the computer. The system will either use services to perform operations or, when a task is too complex to be automated, will ask for human input.

Without the ability to measure, it will be difficult to have any progress. The progress of the measurement systems is the root of progress. The monitoring is to track the main processes, so that the data about their status can be collected and analysed, and the statistics (Big Data, today) on the performance of these processes can be shared among the stakeholders. Figure 12.14 shows an example of a monitoring system of a BHS. Furthermore, this data can be used to work with the main users and suppliers to improve their interactions with these processes.

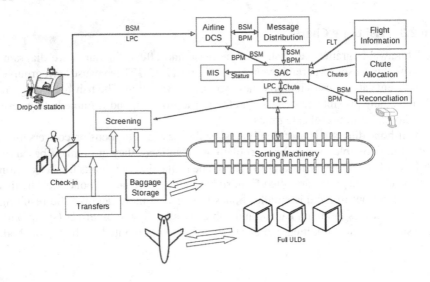

Figure 12.14   Monitoring a BHS system

12.7   PROCESSES FIRST, OPTIMISATION NEXT

The approach presented in this chapter is based on the continuous improvements concept. Business process management is one of the efficient methodologies to handle the continuous implementation of changes and improvement to any business. The analysis and the simulation tools give insight into the results of technology, organisation, and processes changes before implementing any changes. The planning and preparation are vital factors for any activity, and as the quote says *"I you fail to plan, you plan to fail"*.

---

[3]Not all the tasks can be automated. There is the limit, so far, some tasks are complex to be automated.

13 Decision-Aid Techniques

13.1 INTRODUCTION

In single objective optimisation problems, the feasible set is totally ordered per an objective function f. For two solutions s_1 and s_2 one has either $f(s_1) > f(s_2)$ or $f(s_1) \leq f(s_2)$. In contrast, multiple optimisation problems present a set of *optimal* solutions which are quite difficult to order. One of the approaches is that, once the solutions have been evaluated, a vector whose components represent the trade-off in the *decision* search space is produced. Then, the Decision Maker (DM) implicitly chooses an *acceptable* solution by selecting one of these vectors (an acceptable solution!). The decision-maker can be a machine or a human. Dashboards and graphics can be used to visualise the problem at hand as well as the outcome of the decision taken.

In practice, taking decisions under multiple objectives has emerged as an approach to assist in the process of taking decisions which best satisfy a multitude of conflicting objectives. There exist several methodologies for Multi-Criteria Decision Aid (MCDA) problems. These methodologies could be grouped into different categories, such as the form of the model (e.g., linear, non-linear, stochastic, etc.), the features of the decision space (e.g., finite or infinite), or the solution process (e.g. prior specification of preferences or interactive) [56].

There are many MCDA techniques such as decision trees, knowledge-based expert systems, neural network classifiers, etc. In the next sections, a short introduction is given to the most used techniques in MCDA.

13.2 DECISION TREES

The decision tree is a decision tool that uses a tree-like graph or a model of decisions. It is often used in operations research, to identify a strategy that is most likely to achieve a goal. More specifically, it is an explicit representation of all the possible scenarios (paths in the graph) from a given state. Each path corresponds to decisions made by the agent, actions taken, possible observations, state changes, and a final outcome node, etc. The decision tree is also a descriptive means to calculate the conditional probabilities of a system (as shown in Figure 13.1).

The weakness of that method is its rigidity. Indeed, the tree takes several criteria into account; thus, if we add a new criterion or suppress an existing one, the decision-tree should be completely redesigned. The possibility of adding or deleting some criteria is very important in a field where new objectives, rules, technologies, etc. may appear frequently and modify the existing rules.

DOI: 10.1201/9781003432920-13

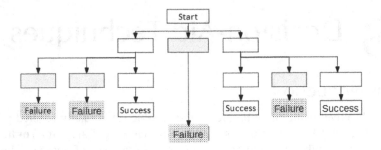

Figure 13.1   Decision tree structure

13.3   KNOWLEDGE-BASED EXPERT SYSTEMS

Experts solve problems using large amounts of specialised knowledge, called domain knowledge using rules of thumb (called heuristics) that they have learnt and refined over years of problem-solving experience in a given domain. The knowledge accumulated is often huge and the experts rapidly narrow down the search by recognising patterns and thus use the appropriate heuristics. Computer programs of Knowledge-Based Expert Systems (KBESs), are quite often used to solve problems in the engineering, mainly the non-structured problems that demand high expertise.

The solution process often involves skilled manipulation of large quantities of knowledge, in a trial and error approach, starting with certain assumptions and revising them when it is necessary till the solution is achieved [169]. Figure 13.2 illustrates a representation of that structure.

Valavanis [181] presented a thorough description of the fundamentals of engineering based expert systems and their knowledge representation techniques. He presented the most important expert system development tools and the existing operational expert systems in many different engineering domains.

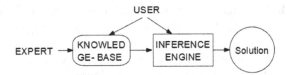

Figure 13.2   Structure of KBES

The two main elements of a KBES are first, a knowledge-base, which is a collection of facts, rules of thumb and causal models of the behaviour of the system. Second, an inference engine, is used to monitor the execution of the program by using the knowledge-base to modify the solution. The inference engine uses the information provided by the user and the rules of the knowledge-base to find a solution.

KBES have produced good results in various areas (design, optimisation, etc.). They are very efficient in the cases where human expertise is well-developed and can be easily defined with simple and sharp rules. But, when the knowledge is imprecise, they lose completely their efficiency.

Often, it is very difficult to predict the effect of the modification of any parameter of the system while the user does not understand how it works. This effect, which is called the "black box effect", is mainly due to the fact that the way of working of the inference engine is out of the user's control. Another weakness of such systems is that they give only one solution, even if other solutions could be possible.

13.4   MULTI CRITERIA DECISION-AID

Scientists from a variety of disciplines have contributed to the development of the field of MCDA methods over the past 40 years. These methods have become a central component of studies in management, economics and industry, etc. They have been, especially designed for multi-criteria problems. Moreover, they can fit the problem better than the conventional approaches. The MCDA aims to give the decision-maker tools for solving decision problems where several – often contradictory – points of view must be considered.

The main characteristics of the classical problems is that generally an optimal solution (which is the best simultaneously from all points of view) that does not exist. These methods do not yield "objectively" the optimal solution, but rather one of the best solutions. Therefore, the focus here is not about the "optimisation methods" but rather "decision-aid methods". A sample of data for such problems can be presented as shown in Table 13.1.

Where $A = \{a_1, a_2, ..., a_n\}$ is the set of the possible solutions of a given problem. $\{f_1, f_2, ..., f_j, ..., f_k\}$ is the set of k criteria used to evaluate these solutions. The value $f_j(a_i)$ is the evaluation of criterion j for solution a_i.

The goal is to find the solution which globally fits the most a set of criteria. For the sake of clarity, we will illustrate the method presented below with an example. Let us consider a simplified example of a factory design and a selection process. The first

Table 13.1
Evaluation matrix

	$f_1(.)$	$f_2(.)$	...	$f_j(.)$	...	$f_k(.)$
a_1	$f_1(a_1)$	$f_2(a_1)$	...	$f_j(a_1)$	...	$f_k(a_1)$
a_2	$f_1(a_2)$	$f_2(a_2)$	...	$f_j(a_2)$	...	$f_k(a_2)$
...						
a_i	$f_1(a_i)$	$f_2(a_i)$	...	$f_j(a_i)$	...	$f_k(a_i)$
...						
a_n	$f_1(a_n)$	$f_2(a_n)$	...	$f_j(a_n)$	...	$f_k(a_n)$
	ϕ_1	ϕ_2		ϕ_j		ϕ_k

Table 13.2

Evaluation table (plant design and selection process)

Id	Cost $(.10^3)$	Capacity	Consumption (kw)	Staff
S_1	10	70	600	10
S_2	15	100	700	7
S_3	20	110	1500	6

task is to identify the criteria which have an influence on the decision to be made. These criteria are generally of different nature: economic, operational, aesthetic, ergonomic, etc. Let us consider that only four criteria are important here:

1. cost, which should be minimised,
2. capacity, which should be maximised,
3. consumption, which should be minimised,
4. staff, which should be minimised.

Having selected the above decision criteria, the evaluation table has to be built for each criterion; to associate an evaluation for each solution. If the set of possible solutions = $\{S_1, S_2, S_3\}$ is considered. The results are illustrated in Table 13.2.

Given the evaluation table, the dominance relation can be defined as (a, b $\in$ A):

- a dominates b for one criterion f_h
  * if $f_h(a) \geq f_h(b)$ (if criterion f_h has to be maximised) or
  * if $f_h(a) \leq f_h(b)$ (if criterion f_h has to be minimised)
- a dominates b (a D b)
  * if $f_h(a) \geq f_h(b)$ (if criterion f_h has to be maximised) or
  * if $f_h(a) \leq f_h(b)$ (if criterion f_h has to be minimised)

$\forall$ h=1, ... , k (with at least one > (or resp. one <)).

This means that "a" dominates "b" if and only if "a" is better or equivalent to "b" for all the criteria with at least one being strictly better.

One can easily notice in the example above, that such a dominance relation cannot be defined between the three solutions. Indeed, solution S_1 is better than solution S_2 for the *cost* and *consumption* criteria, but solution S_1 is worse than solution S_2 for the *staff* criterion. Indeed, it is obvious that such approach does not generally lead to a complete ranking on the set of solutions.

The problem is not mathematically well-stated and the notion of optimal solution does not exist. However, the problem is most often economically well-stated as it expresses the different and possibly conflicting objectives of the decision–maker.

The MCDA methods have been designed to solve this kind of problems. One can consider three great families in the field of MCDA:

1. Interactive methods,
2. Complete aggregation methods,
3. Outranking methods.

13.4.1 INTERACTIVE METHODS

The first family of these interactive methods consists in alternating calculation and dialogue steps. These methods are mostly developed in the field of multiple objective mathematical programming. Some of them can be applied even to more general cases such as "expert systems", and are useful when the set of the possible alternatives becomes too large, and especially when it is continuous. Vanderpooten [184] presented an interactive method which alternates two kinds of phases:

- Calculation phases executed by an analyst or a computer,
- Dialogue phases actively involving the DM.

At each iteration, the procedure presents to the DM a proposal which consists of one or several alternatives against which the DM reacts and provides preference information (dialogue phase). Then, this information is used by the procedure to construct a new proposal (calculation phase). After a certain number of iterations, a final prescription is derived from the last proposal.

13.4.2 COMPLETE AGGREGATION METHODS

This family of methods, which are called "complete aggregation methods", aggregate the different points of view into a unique function which must subsequently be optimised [155]. The first idea to evaluate an alternative with respect to the other ones, is to sum the scores for all the criteria. One can also associate a weight to each criterion and sum the scores multiplied by the weight associated with each criterion. This method is called "the weighted sum of the scores". Let us call E_i this evaluation for the alternative "i". We have the following utility function:

$$E_i = \frac{1}{W} \sum_j f_j(a_i) w_j$$

where $f_j(a_i)$ is the evaluation of alternative "a_i" for criterion j, w_j is the weight associated with criterion j, and W is the sum of the weights of all the criteria.

This method is quite powerful when you have a number of good alternatives (solutions) to choose from (and there is no clear and obvious preferred solution), and diverse factors to take into account to select the best solution.

To use this method, let's put the options as rows on a table and on the columns, put the factors one needs to consider. Then, give a score (between 0 "poor" to 10 "very good") to each combination (choice, factor), and then assign weights to show the importance of each of these factors. Thereafter, multiply each score by the weight of its corresponding factor, to show its contribution to the overall selection. Finally,

Table 13.3

Example of compensation of a bad score by a good one

Criteria	1	2	3	4	5	6	E_i
Action 1	2	10	9	8	9	8	46
Action 2	6	8	8	7	8	7	44

sum up the scores for each option, and then the highest scoring option should be the best option (solution!).

This type of evaluation technique has two major shortcomings, that were identified by Scharlig [153, 155]. The first one is that those methods hide the unusable character of an alternative, because it allows the compensation of a bad score by a good one. This is the case in the example shown in Table 13.3 for a simple summation (scores from 0 to 10 and weights equal to 1). Action 1 receives a better evaluation than action 2, while being almost unusable for criterion 1. A good method should of course reject immediately action 1. The second shortcoming is more serious because it relies on the fact that a unique judge evaluating an action for different criteria does not generally use the same scale for each of them. It is thus the same as having different judges for the evaluation of the same alternative.

Let us consider an example, illustrated by Figure 13.3, where two judges should evaluate the same action. They will generally agree on the maximal and the minimal score, but between those two extremities, their scale may have a different evolution.

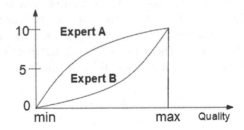

Figure 13.3   An example of evaluation scales for the same action by two experts

This means that when one does the summation of the scores —with or without weights —one adds incomparable units (we should compare apples with apples!). However, it does not mean that the experts do not work methodically. The ranking they give will probably generally be correct, but the method consists of adding the evaluations that can lead to mistakes [27, 185]. Thus, the out-ranking methods are introduced to overcome this kind of problem (see Mareschal [27]).

13.4.3 THE OUTRANKING METHODS

The two most known methods in that field are ELECTRE (which stands for ELimination and Choice Expressing Reality) [148] and PROMETHEE (which stands for Preference Ranking Organisation METhod for Enrichment Evaluations) [27]. The idea behind these methods is that it is better to accept a result which is less rich that the one yielded by the complete aggregation methods. Thus, it is better to avoid mathematical hypotheses which are too strong and demanding too complex questions to the decision-maker. In these methods, no more utility functions are used.

In ELECTRE, the aim is to obtain a subset N of actions such that any action which is not in N is outranked by at least one action of N. The latter subset (which be made as small as possible) is thus not the set of good actions, but it is the set in which the *best compromise* can certainly be found.

The PROMETHEE was developed by Vincke [185], Mareschal [27] and J.P. Brans [26]. The method includes three steps:

Step 1: Enrichment of the preference structure

The notion of generalised criteria is introduced to consider the amplitudes of the deviations between the evaluations. For this purpose, the preference function P(a, b) gives the degree of preference of a over b for criterion f. In most cases, we can assume that P(a, b) is a function of the deviation d = f(a)-f(b) [27]. We consider :

- $0 \leq P(a, b) \leq 1$
- P(a, b) = 0 if d$\leq$0, no preference or indifference
- P(a, b) $\approx$ 0 if d > 0, weak preference
- P(a, b) $\approx$ 1 if d >> 0, strong preference
- P(a, b) = 1 if d >>>0, strict preference.

P has to be a non-decreasing function of d, with a shape like that of Figure 13.4.

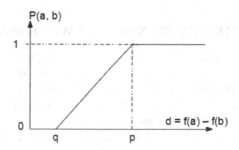

Figure 13.4   An example of preference function

The generalised criterion associated with f(.) is then defined by the pair (f(.), P(., .)). The generalised criterion is associated with each criterion f_j, j=1, ..., k.

Step 2: Enrichment of the dominance relation

A valued outranking relation is built to consider all the criteria. Let us suppose that a generalised criterion (f_j(.), P_j(., .)) has been associated with each criterion

$f_j(.)$. A multi-criteria preference index $\pi(a, b)$ of "a" over "b" can then be defined taking into account all the criteria:

$$\pi(a,b) = \sum_{j=1}^{k} w_j . P_j(a,b), \text{ where } \left(\sum_{j=1}^{k} w_j = 1 \right)$$

where $w_j > 0$ (j=1, ..., k) are weights associated with each criterion and are positive numbers that do not depend on the scales of the criteria. It is interesting to note that if all weights are equal, $\pi(a, b)$ is simply the arithmetic average of all the $P_j(a, b)$ degrees (j=1, ..., k). $\pi(a, b)$ expresses how and with which degree "a" is preferred to "b", and $\pi(b, a)$ how "b" is preferred to "a", over all the criteria. For each pair of alternatives, a, b $\in A$ the values $\pi(a, b)$ and $\pi(b, a)$ are computed. In this way, a complete valued outranking relation is constructed on A.

Step 3: Exploitation for decision aid

Let us consider how each alternative a$\in$A is facing the *n-1* other ones and therefore define the two following outranking flows:

- the positive outranking flow: $\phi^+(a) = \frac{1}{n-1} \sum_{\substack{b \in A \\ b \neq a}} \pi(a,b)$

- the negative outranking flow: $\phi^-(a) = \frac{1}{n-1} \sum_{\substack{b \in A \\ b \neq a}} \pi(b,a)$

The positive outranking flow expresses how much each alternative solution is outranking all the others. The higher $\phi^+(a)$, the better the alternative. $\phi^+(a)$ represents the power of a, its outranking character. The negative outranking flow expresses how much each alternative is outranked by all the others. The smaller $\phi^-(a)$, the better the alternative. $\phi^-(a)$ represents the weakness of a, its outranked character [4, 27].

13.5 MULTI-CRITERIA DECISION AID – A WAY FORWARD

It is easy to justify the choice of the MCDA method as opposed to the conventional approaches such as decision trees, knowledge-based expert systems, etc. The decision trees technique presents the major inconvenience of being very rigid, even if it is able to take different kinds of criteria into account. The knowledge-based expert systems are very efficient in fields where human expertise is very sharp and easy to model. When it is not the case, they lose a large part of their efficiency leading to a very large set of rules and to a system which is very difficult to control. Another weakness of these systems is that they give only one solution, even if another solution would have been possible.

But the major reason for the weaknesses of these methods is simply that they do not really fit to the problems that are considered in this chapter. Indeed, the intrinsic nature of the design problems is multi-criteria. We have different criteria influencing the decisions, and these criteria have different natures and units. Moreover, they are often contradictory, and the way they influence the decision is rarely sharply defined.

14 Optimisation and Artificial Intelligence

14.1 INTRODUCTION

Design is the process of specifying a description of an object (e.g., product, program, service, etc.) that satisfies a collection of constraints. The aim is to find a solution (the best one!) that satisfies different constraints. In general, industrial problems are characterised by a infinite number of feasible solutions. For small problems, the optimal (best) solution can be found by an enumeration method. For big problems, the trend is to use optimisation techniques *heuristics* rather than exact methods, as these enumerative methods are time's greedy.

Search algorithms define a design problem in terms of a search problem where the search space is a space filled with a set of points. Each point in that space defines a *solution*. The design problem is the problem of searching for the best solutions somewhere in the space of valid ones. The whole procedure is composed of three steps: define the problem, fix the goal, and finally, use a method to reach this goal.

Optimisation is everywhere, it is a part of human nature to strive for goals and to optimise the actions towards these goals. Problems can be represented with models and formulated in the language of mathematics. A mathematical model is always a simplified representation of a true physical object that is being investigated. Mathematics and physics describe the world around us in terms of physical concepts and theoretical objects. Optimisation problems are expressed as the minimisation of a cost function that depends on a set of input parameters.

In many optimisation applications, the researchers are more interested in finding a "good enough" solution than proving the optimality. "Good enough" may of course mean "very good, and much better than would be found by other means" [152].

So far, BHS many designers still have some lack of knowledge in the field of optimisation, AI, MCDA, etc. The purpose of this chapter is to bridge the gap between the AI specialists and the BHS designers.

14.2 ARTIFICIAL INTELLIGENCE

The Artificial Intelligence (AI) techniques are iterative procedures (meta-heuristics) that combine different strategies based on computerised models in order to obtain high-quality solutions to complex problems. They are concerned with the development of computer programs that emulate the intelligence of humans. The well-known metaheuristics that have been successfully applied to optimise real-life applications are simulated annealing, tabu search, ant colony optimisation, genetic algorithms, and Artificial Neural network, etc. These meta-heuristics are inspired, by the physical

DOI: 10.1201/9781003432920-14

annealing process, the proper use of memory structures, the observation of real ant colonies and the Darwinian evolutionary process [136, 151].

With the era of Big Data, reappeared the need to solve optimisation problems of unprecedented big sizes. Machine learning, social network science and computational biology are examples of application domains where it is possible to formulate the problem with millions of variables. Classical optimisation algorithms are not designed to deal with instances of this size; new approaches are needed. Thus, the challenge is to come up with optimisation algorithms capable of working in the big data setting.

14.3 OPTIMISATION PROBLEMS

Lawler [104] defined Combinatorial Optimisation Problems (COP) as follows: "Combinatorial Optimisation is the mathematical study of finding an optimal arrangement, grouping, ordering, or selection of discrete objects". The application of the optimisation techniques to industry can be classified into many categories: transportation, logistics, supply chain management, manufacturing, telecommunications, biology, electrical power systems, electronics, military, etc.

Any industrial problem can be turned into one of these problems: operations sequencing and scheduling [139, 141], grouping [52, 123], routing [179], graph colouring [107], partitioning [7], assignment [29], etc.

In the field of airport business, the runway planning and gate assignment for aircraft [10, 45, 115] and check–in scheduling, passengers' boarding and de-boarding, etc. are some examples of the optimisation problems.

The aim of a COP is to search and determine the most suitable solution for optimising (minimising or maximising) a set of feasible solutions to real life problem. For example, in mechanical engineering, an engineer wishes to design a car consisting of composite materials. The engineer aims to determine what is the most suitable shape to bond the various piles of composites in order to maximise the strength of the car. The engineer, using optimisation tools, will conceivably design a lighter, stronger, attractive and safer composite car.

In the airport baggage handling, a designer wants to design a schedule taking in consideration components of the site-work and functions such as time, personnel, location of bags, sorting, bags screening, baggage storage, reconciliation, etc. The designer would minimise the time of performing the baggage handling resources schedule. The aim of the optimisation is to find a cheaper and more acceptable schedule and maintain the requirement of BHS consistency. From the above, the BHS problem can be formulated as a COP problem [73]. Some of the most known optimisation problems and categories are introduced in the following sections.

14.3.1 THE TRAVELLING SALESMAN PROBLEM

The Travelling Salesman Problem (TSP) is one of the most studied problems in computational mathematics and it was first formulated mathematically in 1930 [35]. The

challenge is to find the shortest route visiting each member of a collection of locations (e.g., cities, warehouses, etc.) exactly once and returning to the starting point.
The TSP can be modelled as an undirected weighted graph, such that locations are
the graph's vertices, paths are the graph's edges, and a path's distance is the edge's
weight (as shown in Figure 14.1). The objective is to minimise the travelled distance
after having visited all the vertexes exactly once.

The TSP problem is computationally quite difficult to solve optimally, nevertheless, many heuristics and exact algorithms have been published, thus some instances
with millions of locations can be solved completely. It is known as an NP-hard[1]
problem, and it is an important topic in operation research and computer science.

Even in its simplest formulation, the TSP has several applications, such as logistics, planning, routing, etc. In many applications, additional constraints such as
limited resources, time windows, etc. may be added as constraints to the problem.

In the BHS business, the optimal route between the different locations of an automated guided vehicle is an example of the TSP problem as explained in Chapter 6.

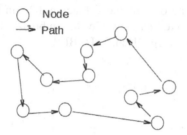

Figure 14.1   The TSP example

14.3.2   THE BIN PACKING PROBLEM

The Bin Packing Problem (BPP) is the problem of packing a set of items into several bins such that the total weight, length, volume, etc. that does not exceed some
maximum values. The 1D-BPP (one dimension) objective is to find the minimum
number k of identical bins of capacity C (positive integer) needed to store a finite
collection of weights w_1, w_2, w_3, ..., w_n. It is not allowed to have bins that have the
items weight's sum exceeding the bin's capacity.

This problem is an NP-Hard problem [57], and in general, finding[2] an exact minimum number of bins takes exponential time.

A simple algorithm (the first-fit algorithm) takes items in the order that they come
and places them in the first bin in which they fit. In 1973, Ullman [86] presented a

---

[1]NP-hard: Non-deterministic Polynomial-time hard problems are problems for which there is no
known polynomial algorithm, so the time to find a solution grows exponentially with problem size. Although it has not been proven that there is no polynomial algorithm for solving NP-hard problems.

[2]Find solutions using enumeration techniques.

new algorithm that could differ from an optimal packing by as much at 70%. The alternative strategy is firstly to order the items from largest to smallest, and then to place them sequentially in the first bin in which they fit. In 1973, Johnson [85] showed that this strategy is never suboptimal by more than 22%, and furthermore that no efficient bin-packing algorithm can be guaranteed to do better than 22%.

The lower bound can be given as:

Min number of bins $\geq$ (Total Weight / Bin Capacity).

Let's look at the following 1D BPP in which there are 6 items and corresponding weights {4, 7, 1, 4, 2, 2} and the bin Capacity C = 10. Thus, the lower bound is "(4 + 7 + 1 + 4 + 2 + 2)/10" = 2. As a result, a minimum 2 bins are needed to contain all the items: the first bin contains {4, 4, 2} and the second bin {7, 1, 2}.

The 2D BPP aims to pack small rectangular shapes into larger rectangles in an optimal way. No two rectangles may intersect or be contained inside one another.

The 3D BPP is the problem where N 3D-boxes are to be packed in a minimum number of containers (bins). No two boxes may intersect or be contained inside one another. The 2D/3D BPP are also NP-hard problem and are used in many industrial applications such as Loading of containers, Placing data on multiple storage disks, Packing advertisements in fixed-length radio station breaks, Layouting, etc. Figures 14.2 shows examples of 1D (resp. 2D and 3D) BPP.

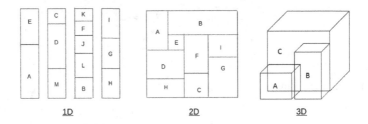

Figure 14.2 Bin packing problems – examples

In the context of the airport business, the layouting or the bin packing problems belong to the field of geometric combinatorial computing. The items to place are the different building, facilities, services, storage, etc. while the building footprint represents the area where to place the different items. The solution generated by the placement should satisfy the different problem's constraints and consider the preferences of the designers.

14.3.3 THE QUADRATIC ASSIGNMENT PROBLEM

The Quadratic Assignment Problem (QAP) aim is to assign a set of departments to a set of locations in order to minimise the total assignment cost [97]. The assignment cost for each pair of departments is a function of the flow between the departments and the distance between their locations [128].

For example, a department location problem is considered with four departments (and four locations). One possible assignment is shown in Figure 14.3 where department B is assigned to location 1, department A is assigned to location 2, department D is assigned to location 3, and department C is assigned to location 3.

This assignment can be written as the permutation $p = \{B, A, D, C\}$. The link between a pair of departments indicates that there is a flow between the departments, and the line's thickness increases with the value of the flow. To calculate the assignment cost of the permutation, the flows between departments (as shown in Table 14.1) and the distances between locations (as shown in Table 14.2) are used. The cost of permutation is given by:

$$cost = \sum distance(i,j) * flow(i,j)$$

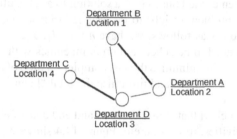

Figure 14.3   The QAP assignment

Assigning gates for flights in order to minimise the walking time for passengers can be modelled as a QAPs problem. Haghani and Chen [74] stated that other versions of the same problems, for example, minimising the amount of baggage that has to be moved or minimising both passenger and baggage movement, can also be formulated as QAPs, etc.

Table 14.1

Flows between departments

Department i	Department j	Flow (i, j)
A	B	3
A	D	2
B	D	1
C	D	4

Table 14.2

Distances between locations

Location i	Location j	Distance (i, j)
1	3	53
2	1	22
2	3	40
3	4	55

14.3.4   THE SCHEDULING PROBLEM

Many research has been carried out to find a solution to a scheduling problem which can be defined as the problem of jobs to be assigned to resources at specific times. The most basic version is as follows: we have n jobs J_1, J_2, ..., J_n of varying processing times, which need to be scheduled on m machines with varying processing capacity. The objective is to minimise the last completion time of a job, minimise the total completion time of all jobs, and minimise the total "lateness" of jobs, etc. and minimise the makespan[3].

Gantt chart, is a bar chart that shows the start and end date of elements in a schedule such as tasks, activities, resources, etc. Figure 14.4 shows a graphical representation of a schedule of four tasks on three machines using the Gantt chart.

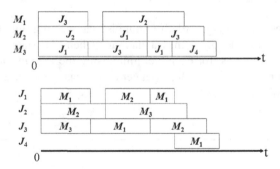

Figure 14.4   Graphical representation (Gantt chart) of a schedule

It is not only the human factor which affects the scheduling problems, but also factors such as the set-up time of machines, cost, and due date, which add further complexity to the scheduling processes. There exist many constraints, and the most known ones are (1) each machine can only process one job at a time. (2) each job can only be processed by one machine at a time. (3) once a machine has started

---

[3]The makespan is the total length of the schedule (that is, when all the jobs have finished processing).

processing a job, it will continue running on that job until the job is completed. Consequently, conventional techniques are not efficient to solve complicated scheduling problems. Scheduling problems can be described by a triplet $\alpha \mid \beta \mid \gamma$.

- α : machine environment (one or two entries)
- β : job characteristics (none or multiple entries)
- γ : objective to be minimised (one entry)

More details about the different scheduling problems can be found in [28, 55]. The resource planning of check-in desks, the gates, the chutes, etc. are examples of scheduling problems (See Chapter 9).

14.4 THE SEARCH APPROACHES

There are at least two types of search approaches, deterministic and non-deterministic ones. A deterministic approach is an algorithm which, for a given inputs, will always produce the same outputs, with the program always passing through the same sequence of states (steps).

Deterministic algorithms (such as integer linear programming, Simplex, etc.) are the most used, since they can be run efficiently on simple machines [193].

At the opposite, non-deterministic (stochastic) search techniques are used to solve intractable discrete global optimisation problems. Their typical domain of applicability are problems of complexity class NP-Complete[4] and beyond. These problems are not perfectly handled by classic deterministic techniques [33].

14.5 THE DETERMINISTIC SEARCH TECHNIQUES

Some of the most known deterministic search techniques will be introduced in the next sections.

14.5.1 THE SIMPLEX TECHNIQUE

A linear programming problem consists of a collection of linear inequalities on several real variables and a fixed linear function which is to be maximised (or minimised). In geometric terms, a closed convex polytope P is considered as shown in Figure 14.5, and defined by intersecting several half-spaces in n-dimensional euclidean space, which lie to one side of a hyperplane [37].

If the problem is linear, the optimum must lie in a vertex (i.e., an intersection of the hyperplanes). To find the optimum, the Simplex algorithm[5] starts at some vertex of

---

[4]NP-complete: Non-deterministic Polynomial time complete. A set or property of computational decision problems which is a subset of NP (i.e. can be solved by a non-deterministic Turing Machine in polynomial time), with the additional property that it is also NP-hard. Thus, a solution for one NP-complete problem would solve all problems in NP.

[5]The method was introduced by Dantzig in 1947 [37].

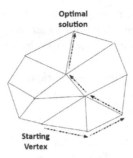

Figure 14.5   Simplex technique – Closed convex polytope

the polytope, and at every iteration chooses an adjacent vertex such that the value of the objective function does not decrease. In general, the Simplex algorithms require a very long execution time for industrial problems.

14.5.2   THE CONSTRAINT PROGRAMMING TECHNIQUE

In Constraint Programming (CP), the problem is defined by a set of variables which can be of any type (continuous or discrete), a (possibly infinite) domain of validity for each of the variables, and a set of constraints (i.e., relations among the variables that must be satisfied) [119]. The CP searches for a set of values for the variables, such that all variables are in their respective domain of validity and all the constraints are satisfied simultaneously.

To speed up the search, the CP uses sophisticated techniques that allow it to use the constraints to restrict the domains of some variables, and to detect the infeasibility (conflict) as soon as possible. The CP is an excellent technique for finding valid (feasible) solutions. However, for optimisation problems, we do not just search for a feasible solution, we search for the *best* one, or at least a high-quality one.

The CP is only useful when the problem is strongly constrained (i.e., when there are relatively few feasible solutions). When there are many (say millions or more) of them, the CP can spend an inordinate time finding the high-quality ones, because is it essentially an *enumerative* approach.

Tarău et al. [177] used the mixed integer linear programming in the Predictive control for baggage handling systems (BHS). The used the method to efficiently compute (sub)optimal routes for destination coded vehicle that transport bags in an airport on a network of tracks.

14.6   THE NON-DETERMINISTIC SEARCH TECHNIQUES

Many industrial problems are known as NP-hard problems. Since these problems have been studied for several decades without yielding an easy method, it is widely believed that no such method exists [33]. Most classical methods (to solve industrial problems) are quite time consuming for large, complex, dynamic and non-linear problems. The size and the complexity of these problems require the development of methods whose efficiency is measured by their ability to find acceptable results within a reasonable amount of time, rather than an ability to guarantee the optimal solution. Industrial problems and many airport's problems are non-linear [44].

As with the stochastic search techniques, these algorithms will not, with probability one, find the global optimum (minimum or maximum), but rather finding a good solution in a reasonable time span. Finding the global optimum usually requires exhaustive search for NP-Complete problems, which is not the primary motivation of sub-optimal techniques like simulated annealing, genetic algorithms , etc.

14.6.1   THE ARTIFICIAL NEURAL NETWORK

An Artificial Neural Network (ANN) is a system, which the main objective is to classify information in the same way the human brain does. The ANN can be taught to recognise, for example, images, shapes, text, etc. and classify them according to the elements they contain. Thus, it can be used to classify data sets into defined classes (classification), or to classify data into undefined categories (clustering), or to guess the future events based on the past ones (prediction).

The ANNs are based on the architecture of the biological neural nets and are composed of the Neurones (nodes) and the Synapses (weights). They consist of input layer and output layers, as well as a couple of hidden layers consisting of units that transform the input into the output layer. The information flow is unidirectional (called feed-forward neural network). Indeed, the data is presented to the input layer, passed on to the different hidden layers, till the output layer. The information is processed and distributed in parallel [76].

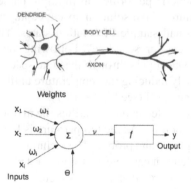

Figure 14.6   Neural Networks (a) vs. ANNs (b)

ANNs have three sets of data. A training set, used to establish the various weights between the nodes. The validation data set, used to fine-tune the network. The test set to check if the ANN can successfully turn the input into the desired output. An ANN works on a system of probability – based on the input data, to make statements, decisions or predictions with a some degree of certainty. The feedback loop enables the "learning" process – by checking whether its decisions are right or wrong, and modifies the steps that takes in the future.

In general, the ANNs are "black boxes", in which the user feeds in data and receives answers. They can fine-tune the answers, but they do not have access to the exact decision making process. They are widely employed for complex tasks such as language recognition, faces detection, pictures classification, fraud detection, text translation, to vehicles driving autonomously on the roads, etc. [1].

14.6.2   THE SIMULATED ANNEALING

The name "Simulated Annealing" (SA) comes from the statistical quantum mechanics (Spin Glasses), where complex systems are brought into their lowest energy state by slowly cooling the system (quasi-statically), allowing the system to make local neighbourhood transitions until a global minimum (of energy) is achieved [93].

The SA is a non-deterministic stochastic search procedure driven by a "temperature parameter" which is a measure of the degree of randomness of the search. The solution space is given a topological structure consisting of local neighbourhoods which themselves are defined by a collection of operators that perform transitions from one state to a ncighbouring state. The transition is being accepted with a probability that is a function of the search temperature.

In a typical annealing algorithm , at high temperature (infinite), all state transitions are accepted with probability 0.5, while at lower temperature the state transitions are accepted per a probability distribution (typically *Boltzmann*[6]). This gives an "algorithmic feel" of pure random search at high temperature, and pure gradient search at low temperature (zero temperature limit).

In practice, the technique is performed in multiple passes. A first pass is used to study the overall structure of the solution space, capturing typical values of the objective function for a small sample of solution states. This information is used to generate a reasonable "cooling schedule". Following the initial pass, the procedure then involves a subsequent pass through the state space, starting the search at a high temperature, and slowly reducing the temperature until a good solution (or near global minimum) is obtained [72] (see Figure 14.7).

At higher temperatures, the technique identifies the major features of the solution space, detecting the major optima. As the temperature is lowered, the finer structure of the solution space is explored and probed. The technique is widely used in industry for the most difficult problems, it can be used for integrated circuit chip design, traffic routing, set partitioning, resource allocation tasks, etc.

---

[6]In statistical mechanics, Boltzmann's equation is a probability equation relating the entropy S of an ideal gas to the quantity W. The Boltzmann formula gives the relationship between the entropy and the number of ways the atoms/molecules of a thermodynamic system can be arranged (annealing).

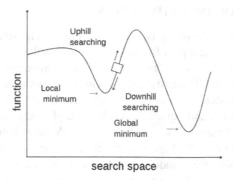

Figure 14.7 Simulated Annealing technique

14.6.3 THE TABU SEARCH

The Tabu Search (TS) technique is a heuristic search to explore the solution space beyond local optimality using adaptive short and long memories [61, 62]. A short-term memory stores the record of the most recent move history and is used to prevent the search cycling in those recently visited solutions. While, a long-term memory is based on the record of the whole move history and is used to guide the search of neighbours of elite solutions (intensification search strategy), or not explored, but promising regions (diversification search strategy) (see Figure 14.8) [196].

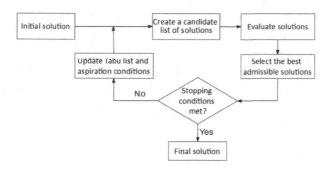

Figure 14.8 Tabu Search technique

The success of the TS techniques can be seen by the improved solutions to problems in scheduling and sequencing with different sizes problems such as flow shop scheduling [131, 176, 190], machine scheduling [61, 102], routing [53, 114], in GPS surveying [73], etc.

14.6.4 THE ANT COLONY OPTIMISATION

The inspiration for the Ant Colony Optimisation (ACO) algorithm came from real ants, that can find the shortest path from a food source to their nest [44]. Ants leave pheromone on the ground while walking, and follow, in probability, pheromone trails that were left by other ants. The first ACO system was introduced by Marco Dorigo, and was called "Ant System". It was initially applied to the TSP, and to the QAP. It is also applied to the BHS and especially routing and planning problems [168].

Technically point of view, when an ant arrives at a decision point, it should decide which way to go (right or left) (Figure 14.9). Since, there is no way for the ants to know which path is best, they randomly pick one. It can be assumed that approximately 50% of the ants will choose the right path, and the other the left path. After a while, the ants who went through the left path, will arrive at the other side, while those who went through the right path will still be on the way. The ants that selected the shorter path will create a strong trail of pheromone faster than those choosing a longer path. This will cause more and more ants to choose the shorter path until eventually all ants have found the shortest path.

The ACO is a population-based algorithm where several artificial ants search for good solutions. Every ant builds up a solution step by step thereby going through several decisions until a solution is found. Ants that found a good solution mark their paths through the decision space by putting some amount of pheromone on the edges of the path. The following ants of the next generation are attracted by the pheromone so that they will search in the search space near good solutions, and so-on [120].

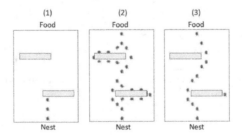

Figure 14.9   Ant Colony

14.6.5 FUZZY LOGIC

The Fuzzy logic method is a computing concept based on the "degrees of truth" rather than the usual "true or false" (binary : 1 or 0) boolean logic which is the basis of the computers. The idea of fuzzy logic was first advanced by Zadeh in 1960s [198]. Zadeh was working on the problem of computer understanding of natural language that is not easily translated into the absolute terms of 0 and 1[7].

---

[7]Whether everything can be described in binary terms is a philosophical question worth pursuing, but in practice much data we might want to feed a computer is in some state in between and so.

It is a type of logic that recognises more than simple true and false values. With fuzzy logic, propositions can be represented with degrees of truthfulness and false-hood. For example, the statement, the dish is sweet, may be 100% true if there is no salt at all, 80% true if there are a few pinches of salt, 50% true if it is half sweet half salted and 0% true if it is salted. Fuzzy logic is how reasoning works, and binary logic is just a special case of it.

In Figure 14.10, the meaning of the expressions *salty, normal, and sweet* is rep-resented by functions mapping a taste scale. A point on that scale has three "truth values" –one for each of the three functions. The vertical line represents a particular-taste that the three arrows (truth values) gauge. Since the dash-dot arrow points to zero, this taste may be interpreted as "not sweet=salty". The round-dot arrow (point-ing at 0.3) may describe it as "slightly sweet" and the dash arrow (pointing at 0.7) "fairly salty" [163].

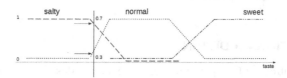

Figure 14.10   Fuzzy Logic

Fuzzy logic seems closer to the way the brains work. Humans aggregate data from different sources and form a set of partial truths which they aggregate further into higher truths which in turn, when certain thresholds are exceeded, cause certain further results and so-on. It has proved to be particularly useful in expert system and other artificial intelligence applications.

14.6.6   PARTICLE SWARM OPTIMISATION

Particle Swarm Optimisation (PSO) is one of the artificial intelligence optimisation techniques that is based on the behaviour of the flock of birds or pool of fishes search-ing for food (as shown in Figure 14.11). The initial work on PSO has been presented by Shi and Eberhart [160].

PSO is a population-based search method where a number of particles are ran-domly generated. These particles move in the multidimensional space searching for the optimal solution. The particles move, according to their velocity, through the multidimensional space taking into consideration the previous individual best posi-tion it had visited and the best position visited by any of the other particles [6, 121]. A particle represents a potential solution of the optimisation problem.

A large inertia weight facilitates a global search while a small inertia weight fa-cilitates a local search. By linearly decreasing the inertia weight from a relatively large value to a small value through the course of the PSO run, the PSO does more global search at the beginning of the run while it does more local search ability near the end of the run. The stopping criteria applied in PSO may be a certain number of iterations, or convergence of particles or if the global best does not change for a specific number of iterations.

Figure 14.11   Swarm phenomenon

14.6.7 EVOLUTIONARY DESIGN

The use of Evolutionary Computation (EC) to generate designs has taken place in many different aspects over the last 50 years. Designers have optimised selected parts of their designs using various ways such as evolution, architects have evolved amazing building plans, computer scientists have evolved the morphologies and control systems of artificial life, etc. [16].

However, research carried out in these application areas tends to be performed by groups isolated within the different fields, and even if there is some cross-disciplinary communication, it seems that the different teams work in ignorance of the results performed by the others. Evolutionary design tries to bring together the different aspects of research under a single heading [16]. Cross-disciplinary research in the EC started many years ago with efforts of Holland [79], Rechenberg [144], Schwefel [158], Fogel [54], Bentley [15, 16], Goldberg [63], Bäck [13], etc.

Evolutionary methods are motivated by Darwin's principles of natural evolution ("survival of the fittest"). A population of solutions to a given problem have to compete to survive. The fitness (how well it solves the problem) of each member of the population is measured. This allows to compare one individual against the others. Thus, the fittest members of the population (this selection is usually done probabilistically), which will survive to the next generations, are selected. The weaker members of the population are *eliminated*. New members are added to the population by breeding from the fittest individuals, and so on.

14.6.7.1 Genetic Algorithm

The Genetic Algorithms (GAs) are stochastic search techniques that are based on the mechanism of natural selection and the natural evolution. The GA, which was introduced by Holland [79], is an optimisation technique inspired by the process of evolution of living organisms. The basic idea is to maintain a population of chromosomes, each chromosome being the encoding (a description or genotype) of a solution (phenotype) of the problem being solved. The worth of each chromosome is measured by its fitness, which is often, simply the value of the objective function in the point of the search space defined by the (decoded) chromosome.

Generally, a *chromosome* is used to represent a solution. It takes a form of a simple string of values called a *gene*. Formally, a gene can be identified as an equivalence relation over the search space. The particular-values that each gene can take are called *alleles*. For example, if the "eye colour" gene can take values "blue", "green", and "brown" then these are its three possible alleles. The position of a gene in its chromosome is its *locus*.

The GAs work in parallel with a number of chromosomes (= solutions). The set of individuals (solutions) of each generation is called a *population*. The chromosomes are characterised by their *fitness* and evolve through successive iterations (*generations*). A population is maintained and the *evolution* plays the role of adaptation of a population to its environment. This adaptation causes the creation of individuals of increasingly higher "fitness". The best solutions are favoured for reproduction every generation, and the offspring are then generated from these fit parents using the crossover and the mutation operators. Thus, evolution drives the population of better individuals (see Figure 14.12).

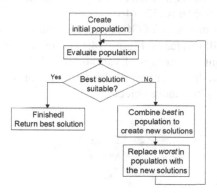

Figure 14.12 Genetic Algorithm steps

The mechanism of the GA is represented by the box "**combine best in population to create new solutions**": this way, better and better solutions are created over the course of iterations of the GA. The meaning of the other boxes is as follows:

1. **Create initial population**: Given the constraints of the problem at hand, several possible solutions are generated.
2. **Evaluate population**: Each of the solutions present in the population is evaluated per the objective function. This assigns a measure of quality to each solution, so that they may be compared in a meaningful way.

3. **Adopt the Best suitable solution**: The GA found a solution that satisfies predefined criteria that qualify it as a solution that may be adopted, or the person that runs the optimiser finds the current best solution of high enough quality to stop the algorithm and adopt the current best solution.
4. **Replace the worst solutions in the population with the new ones**: The worst solutions per the objective function, are discarded and replaced by the new ones created by combining the best ones in the current population.

There are many variations of the standard GA, an exhaustive study can be found in [34, 52, 58, 62, 63].

14.6.7.2 The Grouping GA

Many industrial problems can be easily transformed into grouping problems. Falkenauer [52] pointed out the weaknesses of standard GAs when applied to the grouping problems and introduced the Grouping GA (GGA) to match the structure of the grouping problems. The GGA's operators (crossover, mutation and inversion) are group-oriented aimed to follow the structure of grouping problems.

Like any GA, the GGA achieves optimisation mainly by combining good solutions into new, even better ones. What sets the GGA apart from other GAs, is its special representation of the solutions. Indeed, the GGA explicitly works with *groups* of items rather than the items themselves. This makes the GGA a very powerful technique for *grouping* problems [52, 146].

The genes in the chromosomes of GGA encode groups of objects rather than the objects themselves, as illustrated in Figure 14.13. The objects in the upper part of the figure are grouped into six groups of various sizes. Likewise, the corresponding chromosome depicted in the lower part of this figure features six genes and each of them encoding a group of one more objects.

Figure 14.13 A grouping and the corresponding GGA chromosome

The GGA needs to possess an initial pool of solutions to fill the population. Experience shows that creating an initial population at random is sufficient for the GGA to supply high-quality solutions in a short time. The performance of the GGA can be enhanced with heuristics that make the creation of high-quality solutions easier. Each solution created by the GGA needs to comply with the constraints of the problem.

14.6.7.3 Genetic Programming

Genetic Programming (GP) was developed by Koza in 1992. The GP is a type of Evolutionary Algorithms (EAs) that are typically used to provide good approximate solutions to problems that cannot be solved easily using other techniques. They are used to discover solutions to problems that humans do not know how to solve, directly. According to Koza [100], the EAs (free of human preconceptions) can generate solutions that are comparable to, and often better than the best human efforts.

The GP programs, which can handle tasks like robot motion, are not written in the same way that traditional software is written. They are created automatically, evolved from smaller, simpler programs. Actually, the job of GP programmers is to get a computer to solve a particular problem without telling the computer how to do it. It is, in many ways, the antithesis of the traditional programming, in which human programmers write every command line –telling the computer exactly what to do in every conceivable situation [191].

The GP programmers begin in quite a different manner, by creating the environment in which their programs "evolve". To create that environment, human programmers write software that randomly produces several small chunks of program code. Using the human programmers' description of the problem to be solved, the software then examines these baby programs, and determines which of them come closest to solving it. Figure 14.14 shows a tree that represents a computer program, and the generated program is: 13 * X8 / Y2.

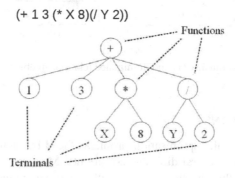

Figure 14.14 Tree to represent a computer program of a GP

14.7 MULTIPLE OBJECTIVE OPTIMISATION PROBLEMS

In general, the search space can be too large and complex to be explored by simple search methods. Multiple Objective Optimisation Problems (MOOPs) involve two "quasi-inseparable" difficulties: search and MCDA (see Chapter 13). The concept of Pareto optimum, which was formulated by Pareto, constitutes the origin of research on MOOPs [134]. A solution s_1 is Pareto optimal if there exists no feasible solution s_2 which would decrease some criterion without causing a simultaneous increase in at least one criterion. There are two classic strategies that were applied with the traditional separation of search and MCDA.

The first one is to make multi-criteria decision to aggregate objectives, then to apply a search method to optimise the resulting figure of merit. The different objectives are combined to form a scalar objective function, usually through a linear combination of the attributes. The approach is well suited to proportional non-competing objectives [46, 126].

The second one, is to conduct the search using different objectives at the same level of importance. A more satisfactory approach is to search for a set of solutions that represents the "best possible trade-off". This leads to a set of alternative solutions and the search phase is followed by making multi-criteria decision to choose among the reduced set. This approach yields the Pareto frontier [134].

By referring to Figure 14.15, O is unique among A, B, C, and D: its corresponding decision vector, $O = (O_1, O_2)$ is not dominated by any other decision vector. That means, O is optimal in the sense that it cannot be improved by any objective without causing a degradation in at least one objective. This is a good practice to integrate MCDA and search and permits to deal with user's preferences [146].

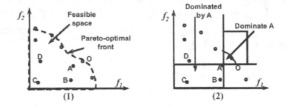

Figure 14.15   Pareto optimality (1) and dominance relations in objective space (2)

14.8   CONCLUSIONS

There is no doubt that the optimisation problems faced by industry (e.g., airports, logistics, etc.) features several different constraints. Most optimisation techniques that are presented in this chapter can cope with a very wide range of possible constraints. The compliance with the problem's constraints is all that is needed for these techniques to solve these complex problems [3, 8, 105].

To give an example, consider various means of locomotion. A mountain bike will not excel on a highway and would probably be left behind all other bicycles behind. But in a forest, a race-bicycle would get stuck in a ravine after a few meters, and an mountain bike would perform much better in those conditions. The means of locomotion must be adapted to the conditions where it is to be used −and the same is true for optimisation algorithms.

Table 14.3 shows non-exhaustive list of possible AI's applications to BHS problems.

Table 14.3

AI potential applications to BHS problems

AI technique	EBS	Resource planning	Routing	AGV	MCS, video coding	Reclaim	APC, batch build
GA		x					
GGA		x					
GP							
SA		x	x				
TS		x	x				
2D BP							x
3D BP							x
TSP			x	x			
Graph colouring		x					
Spanning tree		x					
Simplex		x					
Constraint programming		x					
Swarm			x	x			
Fuzzy logic	x	x					
Ant colony			x	x			
ANN					x		x
MCDA	x				x		

15 Systems Simulation

15.1 INTRODUCTION

The BHS is composed of many individual components, such as sorting machinery, X-Ray machines, check-in, storage, handlers, and so on, and its performance is the outcome of interactions of these different components. It is possible to design experiments to check a BHS behaviour by setting the number of handlers, machines, conveyors, etc. and observe what happens. However, this pragmatic approach to study a system may not be economically feasible.

The question is do the suppliers and the owners know whether (Yes/No) have developed the best solution for a given system, and what process supports the choices made along the way? Those are quite difficult questions, to be answered by Yes or No. Many complex questions arise when an organisation faces an investment decision or a change in its business activities:

- How the proposed system will function?
- How will the revenues and the profits be impacted?
- What effects will this have on operations and staffing?
- What capital investments need to be made and how large will they be?
- etc.

Often, these questions remain unanswered until the change is implemented. By then it is significantly more costly to correct any wrong decisions. A simulation model provides the opportunity to test, change, and develop ideas with computer models, rather than using trial and error approach on the real system. It is also gives insight into the results of the process changes before the changes are implemented.

Most of the BHS life cycle is around 15–20 years years, in which most of the big airports are almost open 20 hours a day, 356 days a year. Any changes to existing infrastructure (e.g., new lines, new equipments, etc.) should be carried out during the 4 hours left of the total 24 hours a day. Thus, the risk in making mistakes is quite big, due to the time limitations and stress. After any change, the whole installation should be be tested on regular basis. Any error may lead to die-backs in the system, which may lead to a complete mess in the airport. Therefore, the simulation and emulation of the existing system, to check what will be the impact of any changes must be analysed before any physical changes. Often, the problems are less expensive and easy to solve in the design process. This enables the handlers to adjust their processes even before the airport commences live operation [112].

The BHS simulation models are essential analysis and communication tools for assisting in the design and the implementation of the heart of any airport terminal, and their ever-growing use is a testament to their effectiveness [80]. Lee et al. [106] presented a simulation tool of airline operations, which helps evaluating the various

DOI: 10.1201/9781003432920-15

Table 15.1
Simulation methods

Method	Field of use	Precision	Description
Global	Planning, feasibility (orders of magnitude)	Weak	Simplified method based on overall ratios and benchmarking of existing BHS. Quick implementation, very few parameters.
Macroscopic	Pre-sizing and sizing	Mean	Method using space allocation ratios and throughputs per module. More complex implementation, high number of parameters.
Microscopic	Design and operation (optimisation)	High	Method by simulating the flow in the BHS. Heavy implementation, very large number of parameters.

airline recovery mechanisms to handle disruptions. Joustra [88] gave some arguments why simulation is necessary in general to evaluate the check-in counters.

There are three main categories of methods to analyse an airport or any system in general. They correspond to different levels of precision and they are adapted to various uses: the simplified global method, the "macroscopic" method and the "microscopic" method by simulation. Table 15.1 presents these three main approaches.

15.2   THE SIMULATION CATEGORIES

15.2.1   THE SIMPLIFIED GLOBAL METHOD

The principle of the simplified global method consists in finding a direct relationship between the surface area of the BHS and the number of passengers that the building can handle (arrival and departure), only as a function of a set of parameters such as the profile of passenger/baggage traffic. The technical and operational characteristics of the BHS are not taken into account in the application of this methodology.

The relationship between overall surface and capacity is built on the basis of ratios derived from the experience of managers, the designers and the specialists from values observed at major airports (benchmark). These rules/methods are refined by reducing the uncertainty ranges thanks to comparative analysis techniques of other BHSs of which the capacity is known and of which the operations are similar to the case we are studying, and adapt the overall ratios with reference to these BHSs.

This rudimentary method which (is in general based on Excel spreadsheets), is quick to implement, makes it possible to evaluate orders of magnitude that can be used in planning on "classic" BHS in the very long term, to compare BHSs with one another, or for feasibility studies (evaluation of a right of way to reserve, economic estimate) but cannot be applied to the sizing, even rough, of a BHS.

15.2.2   THE MACROSCOPIC METHOD

It is a successive application of ratios and formulas to baggage handling equipment in order to calculate the capacity of each module of the system, then the overall capacity of the BHS. This method is more precise than the previous one. It allows a rapid assessment of the capacity of a BHS by requiring a minimum of resources, but if necessary it is possible to refine it by improving the modelling and increasing the number of factors taken into account. It is possible to pre-size a new BHS or a BHS extension on the basis of this method, but it is recommended to validate it by simulations or comparative analyses.

One of the limitations of the macroscopic method is the fact that the passenger/baggage flows (departure or arrival) are not necessarily constant throughout the journey to. The flow of passengers does not flow continuously and regularly throughout the terminal.

The macroscopic method is not able to detect certain phenomena over short periods of time. For example, if the flow of baggage presenting to the screening processes undergoes a sharp but short increase, this can cause a degradation of the quality of service which is not detected by the macroscopic method. The flow of passengers/baggage processed by a module is conditioned by the flow of the modules located upstream. It is therefore recommended to analyse the mode of presentation of passengers/baggage to each module, based on the operation of the BHS, to refine the estimate of the flows and as a result the capacities. However, some BHSs can be such complex systems that without recourse to simulation methods, it is impossible to take into account all the flows and their interactions in a rigorous way.

15.2.3   THE MICROSCOPIC METHOD

The "microscopic" method is based on a simulation approach. It involves modelling the entire system, then creating flows of passengers and baggage. Statistical laws are used to simulate the behaviour of passengers/baggage in the different areas and their processing. It consists in placing oneself in an observer position, in a specific infrastructure, operated according to known procedures. It is the flow of baggage under "borderline" conditions that indicates the capacity of the BHS.

A flow of baggage is generated, we observe how it is passed through the different processing modules. If the quality of service conditions drop below a certain threshold, or if the number of devices is no longer sufficient to handle the flow, saturation is reached: the capacity of the system is exceeded. Otherwise, we repeat with a larger flow of baggage until saturation is reached.

It is difficult to master and anticipate the reaction of the system to certain experimental conditions. Requiring a human to make these decisions can generate inaccurate solutions. In addition, this process can take an unacceptable amount of time. For this reason, in general, the simulation is considered. It starts by developing a mathematical model that resembles as closely as possible the real-life decision situation, and then use a computer-model to solve the problem under various decisions.

Simulation is defined as: *"the imitation of a real-world process or system over time. Simulation involves the generation of an artificial history of the system and the observation of that artificial history to draw inferences concerning the operating characteristics of the real system being represented"* [12].

The simulation technique can be applied to any system whether suitable or not for analytical studies, whether deterministic or random phenomena, etc. In general, a simulation model produces a large volume of results that need to be summarised for interpretation. These results are statistical in nature, and can therefore be interpreted equivocally. Despite these issues, simulation remains a privileged tool in the study of the behaviour of complex systems [171, 172].

15.3   THE BHS SIMULATION

The dynamic discrete event simulation is the method that is used to create mathematical models that are capable of representing the behaviour of a wide variety of complex systems – including factories, airports, etc. Simulation models are data driven (i.e., they can incorporate realistic randomness, they are flexible, they can be visually accurate to assist in understanding any system, etc.).

For BHS, the model contains the equivalent of a network of possible paths that connect the different resources such as conveyors, screening machines, bag storage, operators, etc. The model progresses because of a sequence of decisions and/or rules being executed by the arrival of bags, aircraft, vehicles and scheduled events. The models may include breakdowns and other unscheduled events to test the robustness of the system, and its eventual recovery. Model flexibility is essential to enable users to modify layouts or equipment choices quickly and then generate results to demonstrate which solution works best under certain conditions.

Often, the main objectives of the simulation model are :

- To give confidence that the performance of the designed system will meet the requirements.
- To optimise system performance (what if scenarios).
- To investigate and determine the designed systems' sensitivity to different operating modes and failures.
- To visualise the operational processes of the system.
- etc.

15.4   THE SIMULATION STEPS

In general, a system is composed of a set of individual elements, and its performance is the outcome of interactions between all these elements. The performance evaluation involves generally two steps: (1) mathematical model and (2) model solution. Due to the large number of these elements it is difficult to find a simple model to describe complex systems. Thus, the purpose of the simulation is essentially to develop a mathematical model that resembles as closely as possible to the real-life decision

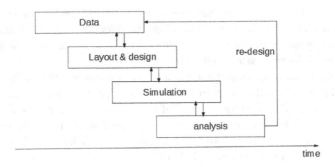

Figure 15.1   Steps to design/simulate systems

situation, and then using a computer-model to solve the problem under various decision circumstances.

Each subsystem is a network made up of three types of elements:

- Links: the role is to allow the movements without modifying its structure. These are spaces purely dedicated to transport (conveyors, lifts, etc.),
- Buffers: whose role is to store the elements of a flow: these are waiting areas (boarding lounges, queues at check-in, baggage storage elements, etc.),
- Processors: which perform a transformation, modifying the characteristics of the flow by means of a control or a transaction (chutes, MCS, etc.).

Figure 15.1 shows a schematic of the process for creating a simulation model. The process involves a set of steps through which one must iterate until a satisfactory solution is obtained. The system under study is first modelled to create a conceptual model of the system which is transformed into a computer simulation. The system's inputs are analysed to create a mathematical representation that will be fed into the simulation model. Then, the outputs of the computer simulation are compared against the outputs (results) of the real system if there exist (or to the designer desiderata if any). If the simulation results do not match with the reality, then corrections must be made either to the simulation model or the input data to match the real system (or designer's desiderata). At each stage, assumptions are made and this process is repeated until a consistent solution is found.

15.4.1   DATA COLLECTION

Although it is not difficult to develop a logical procedure to simulate a system, the difficulties may come from several different sources. Information may not be available; therefore, we should deal with unknowns. Among all the available information, some items are subject to changes, stochastic phenomena are inevitable, etc. Table 15.2 presents some Data for BHS simulation models.

Table 15.2
Some data for BHS simulation models

Topics	Description
Flights	- flight scheduling (planning)
	- number of segregation per flight
Baggage	- plane filling rate (%) (per plane, per category (small/big plane))
	- number of baggage per passenger (per plane, per category (small/big plane))
	- passengers arrival law (constant, normal low)
	- percentage of transfer baggage
	- statistics about bags entering the system (bag profiles: see Chapter 4)
Layout	- number of check-in desks (departure)
	- number of carousels (build area) / MCS / chutes / MUPs
	- number of lanes (or zones) for the baggage storage system
Flows	- number of passengers per year
	- transfer baggage rate (%)
	- number of departure bags/ arrival bags / transfer bags
	- peak flow of local bags / arriving bags / transfer bags
Processing	- check-in occupation rates
	- manual encoding / transfer input stations / terminating input stations
	- sortation/laterals make-up units
	- equipment features (X-ray machines, conveyors speed, ATR read rates)

15.4.2 BASE MODEL DESIGN AND DEVELOPMENT

First, the mechanical drawings are made using 3D design software, then the drawings are transferred to a simulation tool (see section 15.5). Next, the simulation model is developed.

The first stage of the development is to implement all the decision points. A decision point is a point where a decision must be made concerning the destination and/or the status of an entity (exp. bag, ULD, etc.). In the next stage of the development process, the mechanical characteristics of each element are set. The model must show the precise physical behaviour as predicted or desired. Once the model is mechanically defined, the logical system (traffic, capacities, etc.), and the rules are implemented.

During the design, one must deal with elementary systems (called entities) which interact in complex ways to accomplish some common logical end. While it may be easy to identify the set of actions each entity will undertake when faced with certain situations, it is often difficult to determine the collective outcome of the actions of the different entities. At the most basic level, a system is made of entities, activities, resources, paths, and controls. These elements define who, what, where, when, and how the system should behave.

The main elements are given below:

- Entities: the items, which are moving between areas (exp. bags, ULDs).
- Activities: the processes, destinations of an entity leaving a location.
- Resources: the elements that execute a task (exp. operators, conveyors).
- Path Network: the paths imposed to the resources and entities in the system.
- Logic: the rules, the schedule, the capacities of the different elements.
- Outcomes: the expected results data set (exp. flows, capacities, size).
- Scenarios: the different scenarios of input/output (bottlenecks, fall-back modes, failures, cases).

15.4.3   VALIDATION & TESTING OF THE MODEL

The model should be tested according to the following phases :

- Validity: a model is said to be *valid* if it is an accurate representation of the studied system.
- Verification: determines if the conceptual model has been correctly *translated* into a simulation model.
- Credibility: a simulation model is called *credible* if the underlying conceptual model is accepted as valid, the computer program is verified and the results are accepted.

In addition to testing the different alternate scenarios, it is required to test a variety of "what if scenarios" to determine the impact of specific decisions. These "what if scenarios" might include conditions such as failure or non-availability of components, storage loops, belt conveyors, ULD dock stations, or modified staffing levels/productivity, etc. The objective of testing the different "what if scenarios" is to identify the resilience of the system to a variety of operational conditions which might be faced in practice.

All the identified scenarios either during the development of the base model or at a later state, are tested at same level of detail as that defined for testing of the base simulation model. In general, the scenarios only show the influence of one parameter at a time to one of the previous models/scenarios. This enables a better understanding of the impact of each adjustment/change.

15.4.4   OUTPUTS ANALYSIS

Generally, the models are classified per the purpose they serve. Ideally, the aim is to build a model of a system, which describes not only the system behaviour but also selects the "best" set of inputs to obtain certain behaviour from this system. Simulating a system is more than sitting down in front of a computer and drawing a picture of the system. While most simulation software provide tools to assist in the process of model development, a methodology and output analysis is also required. Once a simulation model is verified and validated, experiments can be run.

Often, a lot of efforts are put into developing a system that is the best for certain situations. Even though an efficient design may reduce the number of runs, the amount of output to analyse may still be large. Simulation model has the potential of generating copious amounts of output data. Designers must have skills to analyse these results and synthesise them into a form that is presentable to the decision-makers. As one simulation expert best put it *"The purpose of analysis is insight, not numbers"*.

The model (its assumptions, states, inputs and outputs) is the first outcome of the simulation steps. Next, the following resources should be utilised:

- System expert opinions.
- Inputs from all the stakeholders (exp. customers/managers/users/etc.).
- Observations of the system being simulated or one which is similar.
- Existing theory, particularly analytic results for existing similar systems.
- Relevant results from previous simulations of similar systems.
- etc.

Below some results of a simulation model:

- Occupation of the different resources
- Number of items proceeded (max, min, average)
- Total time spent in the system (max, min, average)
- Number of items passed throughout each resource (max, min, average)
- Number of breakdown of resources
- Number of items transferred between two areas (flows)
- Resources and/or area utilization
- Current and total number of items in the system
- % of missed/late items vs. time/day
- % of handled items by time
- Work in process vs. time
- etc.

15.5 THE SIMULATION TOOLS

There are different kinds for simulation:

- Excel spreadsheet
- Building blocks simulation (macro level)
- Monte Carlo simulation (stochastic, extrapolation)
- 3D animated simulation (micro level)
- etc.

15.5.1 THE MONTE-CARLO SIMULATION

Monte Carlo simulation is a method of estimating a numerical quantity that uses random numbers [49]. The simulation of a bank loan, for example, is not a Monte-Carlo simulation because it involves exact calculations based on the number of monthly payments and the interest rate; no random phenomenon is involved in the calculations. In general, it can be used to determine the reliability of a system, its availability or its average time to failure.

The Monte Carlo simulation is a versatile method for analysing the behaviour of activities, processes, etc. that involves *uncertainty* [149, 154]. It can be used to understand the impact of these uncertainties (e.g., variable demand, variation in processes, weather effects, etc.) and develop plans to mitigate and/or cope with these risks. The objective is to find out what will be the system behaviour in the long term.

The Monte Carlo simulation tool attempts to follow the "time dependence" of a model for which the changes do not proceed in some rigorously predefined fashion, but rather in a stochastic manner (like the passengers demands) which depends on a sequence of random numbers [103]. Its output is not a single value – but a probability distribution of all variables (see Figure 15.2).

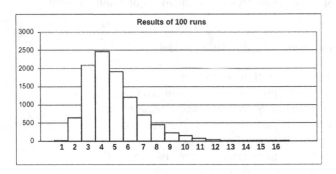

Figure 15.2   Simulation output results (sample)

For a given problem, the Monte Carlo model is developed so that the quantity to be sought is expressed as the expectation of a random[1] variable X, noted E(X). In some cases, the quantity to be calculated is naturally an expectation, for example the average of the number of customers in a queue when the arrival and processing times are random. But the method can also be used to estimate a purely deterministic quantity, for example a surface or an integral, by artificially constructing a random variable to come down to the calculation of an average.

Any problem to be solved by the Monte-Carlo method must be modelled by calculating the mean of a random variable X. In general, a Monte-Carlo model to estimate the mean of a random variable X, has the following steps:

1. Ask for the number n of points to generate;
2. Initialise a *sum* counter to 0;

---

[1]A random variable is the result of an experiment subjected to chance; its expectation is roughly what one would expect to find on average if one repeats the experience a large number of times.

3. For i from 1 to n:
- generate an X_i copy of X
- add the value X_i to the *sum*

4. The estimator is *sum*$/n$, the average of the generated values.

15.5.2 THE 3D SIMULATION SOFTWARE

These software packages have 3D graphical animations and they include a CAD editor that supports accurate 3D modelling. Drawings can also be imported from CAD systems. Once the objects and graphics have been created, the system logic is added to describe the behaviour of the entities using programming languages. The performance-statistics of the simulated system can be displayed using tables or graphics, while the model is running, or after the simulation is completed.

Numerous simulations tools can be used to simulate BHS. Figure 15.3 shows a screenshoot of a simulation model developed under Automod[2] software [11].

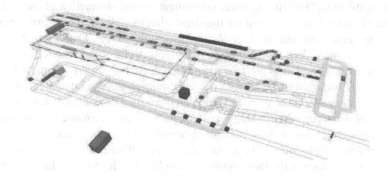

Figure 15.3 The simulation model of a BHS under AutoMod software

15.6 THE INDUSTRY 4.0 AND SIMULATION

According to the proverb "*A fool with a tool is still a fool*", the simulation software are only tools, still to define, to simulate and analyse the obtained behaviour, results, etc. In general, simulation is a time-consuming process. Thus, some elements must be kept in mind while developing a model. There is no need for having one-to-one correspondence between the elements of the real-world system and the simulation model. The model should be specific to the system to study. The objectives and the scope of the model must be defined before the coding begins, and the model should

---

[2]AutoMod is a user-friendly, detailed 3D-animation, efficient simulation and analysis software. Auto-Mod offers a variety of possible applications due to its very efficient simulation core. It can be used to modelling manufacturing processes through warehouse simulation and supply chain simulations, baggage handling systems, etc.

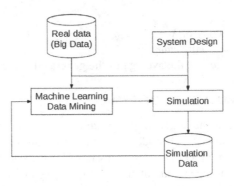

Figure 15.4 Simulation closed loop

be adequate for the determined purpose (i.e., use only as much detail as is essential). In addition, the model building requires simplification and abstraction of the system.

Big data can be used as an input for the simulation models, which in turn generate results of systems to be modified, and/or to be developed, etc. Data mining can be used to analyse these data. Thus, this way the loop is closed as depicted in Figure 15.4. Big data and simulation models, together, can enable airports to make robust, consistent and quick business decisions. Direct comparisons can be made to the current design system operation; to understand and quantify the benefits of any changes. This is useful when considering long-term investments, and to ensure that the efforts are concentrated on achieving the maximum benefit for the stakeholders.

The simulation techniques have a long history and close links with industry, and have been improved since the third industrial revolution. This trend is likely to grow within Industry 4.0 which aims to achieve complete digitisation of manufacturing and many other businesses including airports.

16 The Big Data

16.1 INTRODUCTION

Big data means a large volume of structured and unstructured data –that inundates most businesses. The amount of data that is being stored is huge, and it keeps growing. The other big change is in the kind of data to analyse. It used to be that data fit into tables and spreadsheets, things like the number of customers that came through the airport, etc. Nowadays, the data analysts can also look at "unstructured" data like photos, tweets, voice recordings, sensor data, etc. It is not the amount of data that is important, it is what can be done with the data that matters. The Big data can be analysed for insights that lead to better decisions and strategic moves.

Nevertheless, not much of data collected is actually being analysed and exploited. Most ends up dormant and unused. One can take data from any source, in order to analyse it, to find answers that enable cost savings, time reductions, and smart decision–making, etc. The act of gathering and storing large amounts of information for eventual analysis started a long time ago. What is new, is that the concept has gained insight early 2000s [118]. The recent technological advances, the Internet, cloud computing, and the ability to store and analyse data, that have changed the quantity of data we can collect. Thus, today, everything we do is leaving a digital trace (or data), which we (and others) can use and analyse.

The big data can be articulated on (Figure 16.1) [142]:

- Volume: Organisations collect data from a variety of sources, including social media, data from sensors, machine-to-machine data, etc.
- Velocity: Data stream in at a high speed and is dealt with in real time. The RFID tags, sensors and the Internet of Things (IoT) are driving the need to deal with tera-bytes of data.
- Variety: Data are received in all types of formats. It can be structured, numeric data, unstructured text documents, email, audio, video, etc.
- Variability: Data can be really inconsistent over time.
- Complexity: Data come from multiple sources, which makes it difficult to link, match, clean and transform data across different systems.

For the question, how to make advantage of using big data, the answer lies in combining advanced modelling techniques with the ability to make sense of a huge amount of operational data. The purpose is to uncover insights, thus, the more knowledge one have, the more confident will be in making decisions. The primary value from big data comes not from the data itself, but from its processing and analysis and the insights and services that emerge from that analysis.

DOI: 10.1201/9781003432920-16

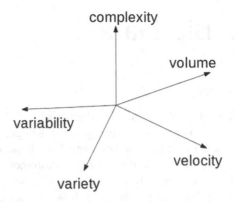

Figure 16.1   Big data fronts

16.2   BHS AND BIG DATA

Big data is the basis for emerging technologies such as predictive analytics and machine learning. For some big airports, over a 100 million log lines are collected daily. By logging data from airport BHSs, we can use full-text searches to gain insights into critical maintenance programmes.

Big data and machine learning are the foundation for analysis which enables airports to react, prevent and predict maintenance needs. Once the data has been crunched by the analysis tools, user-friendly dashboards can be used to visualise the system diagnostics. More, unsupervised machine learning can be used to detect anomalies in system log files. Thus, Big data will also enable the prediction of the MTBF[1] using alarm data. The sensors and devices of the BHS equipment can provide much data, to help predict maintenance issues. Data analytics can also help refine operational decisions [14]. Real-time data can also be used to assist management in its forecasting, planning and reporting functions, such as staff forecasts based on passenger data, etc.

Conventionally, a reporting software is used to get statistics from the SAC and to get traceability of each bag through the system. It gives the necessary information about the different transactions done per bag and is able to print report when necessary. The processes of sorting and tracing bags and re-conciliating them with passenger data might seem simple and obvious at first sight, over the years, the demand of the airports, has seriously evolved and implies to store larger and larger amounts of data.

Today, all is about turning raw data (equipment data, alarm logs, system events, process data and sensor data, etc.) into insights. Thus, enabling the manager (human or a machine) to make decisions and act based upon the results. Thus, the business will move from "what we think and guess" to "what we know".

---

[1]MTBF: Mean Time Between Failures.

The airports expect highly precise and detailed explanations for exceptional cases (baggage lost, wrong sorting, tracking errors, system failures, etc.). This leads to recording huge amounts of data about low-level processes (all messages exchanged with other systems, PLCs, etc.) and about the behaviour of the system and its components (history of CPU, memory and disk usage, low level logs of each process, etc.). To address such requirements using traditional databases and development tools can quickly became useless. Thus, lead to the conclusion that big data tools and technologies are definitively the right answer to such problems.

16.3   APPLICATIONS

One of the issues of BHS is the no-read bags. A pre-trained neural network (see Chapter 14) to implement image recognition using machine learning can be used to tackle this problem. A thousands/millions of images (Big Data) of bag tag reads and no-reads are fed to a neural network are processed through the machine-learning software which instantly recognises whether the bag has a tag or not. The software then selects and sends the best images to the operator for remote (Video Coding: see 2.2.9). More, the machine can not only select images, but also can read and automatically encode the information on the bag tag [70].

Big data is used by various actors of the airport business. Airline to optimise the services they give to their customers, the airports to optimise their operation, etc.

For the airlines:

- Passengers database can be used to determine which new routes could be open by finding final *destinations* of passengers in order to determine if a direct connection between airports should be open, when and how often.
- The types of passengers can help to optimise the kind of *goods or services* to propose to passengers on a plane. Destinations, time schedules also influence the passenger choices.
- etc.

For the airports:

- Data collected about the conveyors can be used to perform predictive *maintenance* for parts.
- Passengers data (first and family names) help to determine the passenger's language. This can be used to optimise the dynamic *advertisement* system of the airport.
- Optimising the path of the passenger in the airport can increase the *airport's revenue*. Passenger able to spend more should have a shorter path in order to give them more time to shop.
- Optimisation of the use of inspection lines depending of the *baggage traffic*.
- Automatic planning of sort destination based on the flight schedule and expected flight occupation, optimisation of early bag storage systems, etc.
- etc.

As in any industry, the airport business has its own priorities to avoid potential disruptions, improving passenger experiences and managing operating costs, etc. Thus, if the big data and analytic tools are combined and used, they can accomplish business-related tasks such as determining the root causes of failures and the defects almost in real-time, etc. The Big data can allow airports to increase the maintenance efficiency of their industrial facilities. By analysing data from their services (e.g., resources, equipment, processes, IT, etc.) airports can identify patterns of activity (e.g., downtimes, productivity, etc.). Thus, they can react on these facts and propose solutions and/or improvements.

The Big data methods can help in creating efficient airports. This can allow to generate fact-based, answers rather than relying on the intuitive, retroactive realisations of traditional research. In the next decade, data will grow by a certain factor (20 or even more). There is an immediate opportunity to learn from the big data successes in other industries. The time to act on big data is coming. More, the newly introduced IATA resolution 753[2] (GDPR[3]) and their traceability requirements will dramatically increased the size of the data stored by the BHS IT systems.

---

[2]IATA 753 resolution is intended to encourage airlines to reduce mishandling by implementing cross-industry tracking for every baggage journey, so that when a bag does not arrive with a passenger there will be much more information available to facilitate recovery.

[3]General Data Protection Regulation

Part IV

Challenges

17 The Epilogue

With cooperation and the exchange of ideas and solutions, Baggage Factory concepts will find their way into practical solutions for the BHS. To study these impacts and to preclude unnecessary (irreversible) changes and a waste of "scarce" resources, the airport stakeholders (e.g., airport, airlines, service providers, etc.) should use tools to evaluate the overall airport process. The criteria to consider in the development of airports, should include the flexibility to accommodate future prompt changes, make compatible the facilities with aircraft characteristics, minimise the passengers cross-flows, allow the traffic peak shaving, minimise the connecting times, guide the passengers, handle the different disasters, etc.

One of the most important key performance indicator (KPI) in the BHS is the average cost per handled bag. Therefore the proposed solutions should focus on enhancing the value of this KPI. Indeed, international airports can handle millions of bags each year, thus a minor reduction in the average baggage handling cost, can lead to quite big savings. Other important KPIs, are the number of mishandled/delayed bags, number of employees, customer satisfaction, etc. The batch build concept, robots, Autonomous vehicles, AI, etc. as opposed to the currently used conventional approaches (push concept, manual handling, etc.), may help to reach these targets.

The airports must be prepared to embrace innovative technologies to ensure a maximum of their efficiency. The tendency is that airports have to offer passengers the option of a 100% self-service journey, where human interaction is eliminated (Is this is really what we are looking for?) thanks to the implementation of automated, self-service technology at every single step of the journey. Staff will be able to focus on providing excellent customer service without needing to perform basic operational processes.

Today, check-in kiosks are at 80% of the world's airports, so investment is now moving to self-service bag drop. Using RFID for bag tracking has the potential to cut mishandling rates more than ever before. Thus, automation and efficiency will go hand-in-hand, with the technology allowing the passengers to walk freely through the airport's processes, from their personal transport to their seats on the plane. We will get to the level where the passengers will have more freedom to do what they want at the airport, whether working, eating, shopping, boarding the plane. Thus, the processes will not dominate the airport journey of the passengers.

In many airports, the loading of baggage into containers is still done manually by operators. More and more semi-automatic or even fully-automated systems for baggage loading are used. A manual station can be equipped with a loading support system (a handling-aid) which relieves the worker from physically hard lifting. The baggage loading at make-up positions is not an exception, and the decision to invest in loading aids is more a question of "when" than "if".

DOI: 10.1201/9781003432920-17

Reclaim on Demand will be the next generation of baggage claim systems. It is an automated system giving passengers visibility into when and where their bags will be delivered to the baggage reclaim area. Rather than collecting baggage from the carousel immediately after arrival, passengers will be able to decide when they want to collect their bags. The passengers will be contacted via an application on their mobile phone, confirming their bag is ready for collection. This will not only reduces the pressure on passengers, but also allows them to spend time in a reclaim shopping area, which provides the airport with additional revenues. It will also reduce the risk to zero of theft or passengers collecting the wrong baggage. This will also reduce wrong reports made by passengers who think their baggage is lost but in fact, has yet to arrive. The 100% traceability of baggage till the reclaim area will reduce the number of missing-bag as well as optimise the cost to recover and reunite passengers and their lost bags.

As the tools to collect and analyse the data (Big data) become more affordable, it will be more and more used in many industries including BHS. Whether you are all for it (the benefits) or against it (worried about big brother is watching you!), it is important to be aware of the technology and try to tune it in to how it will affect your daily life. The knowledge, based on Big data, is a power and the airports will use the data into actionable information to improve the passenger experience and airport operations. By combining the AI and Machine Learning (ML), more effective optimisation procedures that fit the BHS can be obtained.

Although part of contact-less technology at the airport terminal subsystem is already in place, such as drop-off desks and e-gates, another part is still under development and testing. Before the pandemic, there was a plan for technology development and its implementation to be installed in the next 5–7 years so that airports can reach the status to become fully contact-less. Nowadays, after the pandemics years, if an airport wants to continue with its business in a new normal[1] situation, the deadline for the implementation of contact-less technology is counted in months, not in years.

The recent pandemics event has changed completely the known paradigms, new ones should be developed! Thus, airports are at the cusp of a new era, that will be defined less by what is happening in the air than what is happening on the ground through digital transformation. There is no better time to own the future than now. What is clear is no one has any idea about the future. It can drift anywhere! Those who purport they know, they do not tell the truth! Airports do not have the choice to decide, they have to act accordingly!

BHS are not always nice to visit. Indeed, most of time they are full of dust, noisy, too dark, lot of steel and a lot of conveyors, etc. The baggage handling systems would greatly benefit from the implementation of cleaner and more accessible facilities, akin to those found in the automotive industry.

---

[1] A human is a social creature! – by the end of the day, he needs some contact with his fellows.

18 Glossary

6Sigma: Data-driven approach and methodology for eliminating defects
ABB: Automated Batch Build
AGV: Automated Guided Vehicle
AI: Artificial Intelligence
ANN: Artificial Neural Network
APC: Automatic Packing Cell
Automod: 3D Simulation software (www.appliedmaterials.com/)
ATR: Automatic Tag Reader
B&B: Branch and Bound
Baggage Factory: New concepts in Baggage handling systems
Batch Build Area: Area will is Build Lanes, Build Cells, ULD Storage, etc.
Batch Build: Process to load bags into containers (ULDs, CANs, etc.) in batches
BHS: Baggage Handling System
BIDS: Baggage Information Display System
BPM: Baggage Processing Message
BPP: Bin Packing Problem
BRS: Baggage Reconciliation System
BTRS: Baggage Tracking and Reconciliation System
BSM: Baggage Service Message
BSS: Bag Storage System
BSSMS: Bag Storage Management System
Build Cell (BC): Station to load bags into ULD
Build Lanes (BL): lanes to store bags in front of the loading station
BWS: Bag Was Seen in the system
CAN: Container
CAD: Computer-Aided Design
CAE: Computer-Aided Engineering
CE: Concurrent Engineering
CI: Continuous Improvement
COP: Combinatorial Optimisation Problem
CPLEX: Optimisation software package (Commercialised by IBM ILOG)
CT: Computer Tomography
CTX: Computer Tomography X-ray
CUSS: Common Use Self-Service. A type of check-in equipment that allows the PAX to check-in both person and bags
CUTE: Common User Terminal Equipment, equipment making it possible for different airlines to use the same check-in desk
DCS: Departure Control System

DOI: 10.1201/9781003432920-18

DCV: Decision Coded Vehicle
Dispatching: Assigning of jobs to vehicles or vehicles to jobs
DM: Decision–Maker
DP: Dynamic Programming
EDS: Explosive Detection system
ELECTRE: ELimination and Choice Expressing Reality
ES: Evolutionary Strategies
EULD: Empty ULD
FFD: First Fit Decreasing
FIDS: Flight Information Display System
FIMS: Flight Information Management System
FULD: Full ULD
GA: Genetic Algorithm
GDPR: General Data Protection Regulation
GGA: Grouping Genetic Algorithm
Industry 4.0: Fourth Industrial Revolution
I/O Dock: Station to transfer ULDs from/into Tug&Dolly
IATA: International Air Transport Association
ICAO: International Civil Aviation Organisation
IoT: Internet of Things
ISO: International Organisation for Standardisation
JIT: Just In Time
KPI: Key Performance Indicator
Lean: Systematic method for waste minimisation
Load factor: Percentage of occupied aircraft seats on a flight
Loading Cell (LC): Cell to load bags into Pallets (Customised CANs)
LP: Linear Programming
LPC: Licence Plate Code
LPN: Licence Plate Number
MBH: Main Baggage Hall
MBS: Multiple Baggage Storage System
MCDA: Multi-Criteria Decision-Aid
MCS: Manual Coding Station
MCT: Minimum Connection Time
MES: Manual Encoding Station (MCS and MES are synonyms)
MOP: Multiple Objective Problem
MUP: Make-Up Position
OOG: Out Of Gauge
O&M: Operations and Maintenance
PAX: Travel industry standard abbreviation for passenger(s)
PL: Physical layout
PLC: Programmable Logic Controller
PLM: Product Lifecycle Management
PTS: Powered Trolley System

PROMETHEE: Preference ranking organisation Method for Enrichment evaluations
RA: Resource Allocation
RFID: Radio Frequency Identification
ROI: Return In Investment
Routing: Path Planning
RP: Resource Planning
RPC: Robot Packing Cell
SA: Simulated Annealing
SAC: Computer Allocation System
SCADA: Supervision, control and acquisition of data
SITA: Socit Internationale de Tlcommunications Aronautiques
SX: Smoothness Index (SX)
TS: Tabu Search
TVR: Time Variability Ratio
ULD: Unit Load Device
UTL/ATL: Authorised To Load Yes/No
VCS: video-coding system
VSA: Vehicle Storage Area

References

1. C. Lopez-Franco A. Alanis, N. Arana-Daniel. *Artificial neural networks for engineering applications*. St. Louis, MI: Elsevier, 2019.
2. A. Abdelghany and K. Abdelghany. *Modeling applications in the airline industry*. Ashgate Publishing Limited, 2012.
3. A. Abdelghany, K. Abdelghany, and R. Narasimhan. Scheduling baggage-handling facilities in congested airports. *Journal of Air Transport Management*, 12(2):76–81, 2006.
4. G.c Akkaya and C. Uzar. The usage of multiple-criteria decision making techniques on profitability and efficiency : An application of promethee. *International Journal of Economics and Finance Studies*, 5:149–156, 2013.
5. Imad Alsyouf, Uday Kumar, Lubna Al-Ashi, and Muna Al-Hammadi. Improving baggage flow in the baggage handling system at a uae-based airline using lean six sigma tools. *Quality Engineering*, 30(3):432–452, 2018.
6. B.S. Umre Altaf Badar and A.S. Junghare. Study of artificial intelligence optimization techniques applied to active power loss minimization. In *ICAET-2014*, pages 39–45, 2013.
7. M. F. Argüello, T. A. Feo, and O. Goldschmidt. Randomized methods for the number partitioning problem. *Computers & Operations Research*, 23(2):103–111, 1996.
8. A. Ascó, J. A. D. Atkin, and E. K. Burke. An evolutionary algorithm for the over-constrained airport baggage sorting station assignment problem. In Lam Thu Bui, Yew Soon Ong, Nguyen Xuan Hoai, Hisao Ishibuchi, and Ponnuthurai Nagaratnam Suganthan, editors, *Simulated Evolution and Learning*, pages 32–41, Berlin, Heidelberg: Springer, 2012.
9. D. Bailey and E. Wright. *Practical SCADA for industry*. Elsevier, 2003.
10. H. Balakrishnan and B. G. Chandran. Algorithms for scheduling runway operations under constrained position shifting. *Operations Research*, 58(6):1650–1665, 2010.
11. J. Banks. *Getting started with AutoMod*. Bountiful, UT: Brooks-PRI Automation, 2004.
12. J. Banks. *Handbook of simulation: principles, methodology, advances, applications, and practice*. John Wiley & Sons, 1998.
13. T. Bäck. *Evolutionary algorithms in theory and practice: Evolution strategies, evolution programming, genetic algorithms*. New York, NY: Oxford University Press, 1996.
14. M. Bender. 5 ways to reduce costs and improve operational efficiency in baggage handling, 2021.
15. P. J. Bentley. *Generic evolutionary design of solid objects using a genetic algorithm*. PhD thesis, Division of Computing and Control Systems, School of Engineering, University of Huddersfield, 1996.
16. Peter J. Bentley and David W. Corne. Creative evolutionary systems. *The Morgan Kaufmann Series in Artificial Intelligence, 1st Edition*, 2002.
17. Beumer. Automated container handling system. Marketing paper, Beumer Group, 2016.
18. Beumer. Baggage forklift. Marketing paper, Beumer Group, 2016.
19. Beumer. Baggage lifting table. Marketing paper, Beumer Group, 2016.

20. Beumer. Baggage loader. Marketing paper, Beumer Group, 2016.
21. Beumer. Baggage manipulator. Marketing paper, Beumer Group, 2016.
22. Beumer. Fully automated system for unloading bags. Marketing paper, Beumer Group, 2016.
23. G. Blokdyk. *Baggage handling system: 3rd Edition.* CreateSpace Independent Publishing Platform, 2018.
24. P.T.C. Vervoort, H.J.A. Bodewes, and J.M. Van den Goor. Method and installation for transporting goods, as well as a combination of a container and a wheel-supported frame for transporting goods. Patent us6540064 b1 veghel netherlands, Vanderlande Industries Nederland, 1999.
25. V. Bracaglia, T. D'Alfonso, and A. Nastasi. Competition between multiproduct airports. *Economics of Transportation,* 3:270–281, 2014.
26. J.-P. Brans and B. Mareschal. The promcalc & gaia decision support system for multi-criteria decision aid. *Decision Support Systems, North-Holland,* 12:297–310, 1994.
27. J.-P. Brans and B. Mareschal. Promethee : Une méthodologie d'aide à la décision en présence de critères multiples. *Editions de l'Université de Bruxelles (collection Statistique et Mathématiques Appliquées),* 2002.
28. P. Brucker. *Scheduling algorithms.* SpringerVerlag, 2004.
29. R. E. Burkard, M. Dell'Amico, and S. Martello. *Assignment problems.* SIAM, 2009.
30. L. Butcher. Etihad opens home luggage checking service in UAE, 2021.
31. J. C. Rijsenbrij and J. Ottjes. New developments in airport baggage handling systems. *Transportation Planning and Technology,* 30:417–430, 2007.
32. J. Carmona, G. Engels, and A. Kumar. *Business process management forum: BPM forum 2017, Barcelona, Spain, Proceedings.* Lecture Notes in Business Information Processing. Springer International Publishing, 2017.
33. R. Chiong. *Nature-inspired algorithms for optimisation.* Studies in Computational Intelligence. Springer Berlin Heidelberg, 2009.
34. C. A. Coello. A comprehensive survey of evolutionary-based multiobjective optimization techniques. *Knowledge and Information Systems,* 1:269–308, 1998.
35. W. Cook. *In pursuit of the traveling salesman: Mathematics at the limits of computation.* Princeton University Press, 2012.
36. Aimable-Lima D. and Fransen D. Baggage system. Patent wo2009098439 a2 great britain, BAA, 2009.
37. G.B. Dantzig. *Linear programming and extensions.* Landmarks in Physics and Mathematics. Princeton University Press, 1998.
38. P. Buckle, G. David and V. Woods. *Further development of the usability and validity of the Quick Exposure Check (QEC).* Research Report 211. Robens Cente for Health Ergonomics, University of Surrey, Guilford, 2005.
39. R. De Neufville. The baggage system at denver: Prospects and lessons. *Journal of Air Transport Management,* 1(4):229–236, 1994.
40. R. de Neufville and A. Odoni. *Airport systems: Planning, design, and management.* Aviation Week Book. McGraw-Hill, 2002.
41. R. de Neufville, A. Odoni, P. Belobaba, and T. Reynolds. *Airport systems: planning, design and management 2/E.* McGraw-Hill, 2013.
42. H. DeBusk, K. Babski-Reeves, and H. Chander. Preliminary analysis of strongarm ergoskeleton on knee and hip kinematics and user comfort. *Proceedings of the Human Factors and Ergonomics Society Annual Meeting,* 61(1):1346–1350, 2017.

43. E. W. Dijkstra. A note on two problems in connexion with graphs. *Numerical Mathematics*, 1(1):269–271, 1959.
44. M. Dorigo and T. Stützle. *Ant Colony Optimization*. MIT Press, 2004.
45. U. Dorndorf, A. Drexl, Y. Nikulin, and E. Pesch. Flight gate scheduling: State-of-the-art and recent developments. *Omega*, 35(3):326–334, 2007.
46. A. Drexl and Y. Nikulin. Multicriteria airport gate assignment and pareto simulated annealing. *IIE Transactions*, 40(4):385–397, 2008.
47. M. Dumas, M. La Rosa, J. Mendling, and H. A. Reijers. *Fundamentals of Business Process Management*. Berlin: Springer, 2013.
48. M. Ebben. *Logistic control in automated transportation networks*. PhD thesis, 2001.
49. R. Eckhardt, S. Ulam, and J. Von Neumann. The Monte Carlo method. *Los Alamos Science*, 15:131, 1987.
50. D. Edwards. Industrie 4.0: Seven facts to know about the future of manufacturing. 2016.
51. Technical Committee CEN/TC 122 "Ergonomics". Safety of machinery - anthropometric requirements for the design of workstations at machinery. *European Committee for Standardization (CEN)*, 2002.
52. E. Falkenauer. Genetic algorithms and grouping problems. *John Wiley & Sons Inc., Chichester, First Edition*, 1998.
53. C-N Fiechter. A parallel tabu search algorithm for large traveling salesman problems. *Discrete Applied Mathematics*, 51(3):243–267, 1994.
54. L. J. Fogel. Biotechnology: Concepts and applications. *Englewood Cliffs, NJ: Prentice Hall*, 1963.
55. M. Frey. *Models and methods for optimizing baggage handling at airports*. TUM-Bibliothek, 2015.
56. T. Gal, T. Hanne, and T. Stewart, editors. *Advances in multiple criteria decision making*. Dordrecht: Kluwer Academic, 1998.
57. Michael R. Garey and David S. Johnson. *Computers and intractability: A guide to the theory of NP-Completeness*. New York, NY: W. H. Freeman & Co., 1979.
58. M. Gen and R. Cheng. Genetic algorithms & engineering design. *John Wiley & Sons Inc, First Edition, Canada*, 1997.
59. A. Ghobrial, C. F. Daganzo, and T. Kazimi. Baggage claim area congestion at airports: An empirical model of mechanized claim device performance. *Transportation Science*, 16(2):246–260, 1982.
60. A. Gilchrist. *Smart Factories*. Berkeley, CA: Apress, 2016.
61. F. Glover and M. Laguna. Tabu search. Boston, MA: Kluwer Academic Publishers, 1997.
62. F. Glover and G. A. Kochenberher. *Handbook of meta-heuristics*. Springer, 2019.
63. D. E. Goldberg. *Genetic algorithms in search, optimization, and machine learning*. New York, NY: Addison-Wesley, 1989.
64. J.-B. Gotteland and N. Durand. Genetic algorithms applied to airport ground traffic optimization. In *The 2003 Congress on Evolutionary Computation, CEC '03.*, pages 544–551, Vol.1, 2004.
65. T. Andersson Granberg and A. Oquillas Munoz. Developing key performance indicators for airports. In *ENRI Int. Workshop on ATM/CNS. Tokyo, Japan. (EIWAC 2013)*, 2013.
66. G. Greg Brue. Six sigma for managers: 24 lessons to understand and apply six sigma principles in any organization. *McGraw-Hill Professional Education Series*, 2005.

67. D.R. Grigoras and C. Hoede. Design of a baggage handling system. *Department of Applied Mathematics, University of Twente*, 2007.

68. Beumer Group. Airport software suite for baggage handling systems. 2019.

69. Beumer Group. Baggage claim on demand: What's in store for the future? 2021.

70. Beumer Group. Gain insight for high efficiency and throughput. 2021.

71. Y. Gu and C. Chung. Genetic algorithm approach to aircraft gate reassignment problem. *Journal of Transportation Engineering-ASCE*, 125, 1999.

72. P. Dare and H.A. Saleh. Effective heuristics for the gps survey network of malta: Simulated annealing & tabu search techniques. *Journal of Heuristics*, 7 (6):533–549, 2001.

73. P. Dare and H.A. Saleh. Near-optimal design of global positioning system networks using tabu search technique. *Journal of Global Optimization*, 25 (2):183–208, 2003.

74. A. Haghani and M.-C. Chen. Optimizing gate assignments at airport terminals. *Transportation Research Part A: Policy and Practice*, 32(6):437–454, 1998.

75. K. Hallenborg and Y. Demazeau. Decide: Applying multi-agent design and decision logic to a baggage handling system. In Danny Weyns, Sven A. Brueckner, and Yves Demazeau, editors, *Engineering Environment-Mediated Multi-Agent Systems*, pages 148–165, Berlin, Heidelberg, Springer Berlin Heidelberg, 2008.

76. S.O. Haykin. *Neural networks and learning machines*. Pearson Education, 2009.

77. Health and Safety Executive. *Manual handling assessment charts (the MAC tool)*. https://www.hse.gov.uk/pubns/indg383.pdf, 2017.

78. C. Hohmann. *Lean Management: Outils, méthodes, retours d'expériences, questions/réponses*. Les Références. Eyrolles, 2012.

79. J. H. Holland. Adaptation in natural and artificial systems. Ann Arbor, AI: University of Michigan Press, 1975.

80. H. Wai Chun and R.T. Mak. Intelligent resource simulation for an airport check-in counter allocation system. *IEEE Transactions on Systems, Man, and Cybernetics, Part C (Applications and Reviews)*, 29(3):325–335, 1999.

81. R.M. Horonjeff, F.X. McKelvey, W.J. Sproule, and S. Young. *Planning and design of airports, 5th Edition*. McGraw-Hill Education, 2010.

82. HSE. *The health and safety toolbox, how to control risks at work*. Crown, 2014.

83. Lödige Industries. Flexible transportation of baggage and cargo ulds. 2021.

84. W.H. Inmon and Daniel Linstedt. 4.4 - introduction to data vault methodology. In W.H. Inmon and Daniel Linstedt, editors, *Data Architecture: a Primer for the Data Scientist*, pages 155–162, Boston, MA: Morgan Kaufmann, 2015.

85. D. S. Johnson. Fast algorithms for bin packing. *Journal of Computer and System Sciences*, 8(3):272–314, 1974.

86. D. S. Johnson, A. J. Demers, J. D. Ullman, M. R. Garey, and R. L. Graham. Worst-case performance bounds for simple one-dimensional packing algorithms. *SIAM Journal on Computing*, 3(4):299–325, 1974.

87. M. Johnstone, D. Creighton, and S. Nahavandi. Status-based routing in baggage handling systems: Searching verses learning. *Systems, Man, and Cybernetics, Part C: Applications and Reviews, IEEE Transactions on*, 40:189–200, 2010.

88. P. E. Joustra and N. M. Van Dijk. Simulation of check-in at airports. In *Proceedings of the 33nd Conference on Winter Simulation*, WSC '01, pages 1023–1028, Washington, DC, USA, 2001. New York, NY: IEEE Computer Society.

89. T. Fog Justesen, A. Høeg Dohn, and J. Larsen. *Allocation of ground handling resources at Copenhagen airport*. PhD thesis, Technical University of Denmark, 2014.

90. A. Kazda and R.E. Caves. *Airport design and operation*. Emerald Group Publishing Limited, 2015.
91. A. Khosravi, S. Nahavandi, and D. Creighton. Interpreting and modeling baggage handling system as a system of systems. In *Proceedings of the IEEE International Conference on Industrial Technology*, pages 1–6, 2009.
92. M. Kim. Video coding as remote encoding technology in baggage handling. *AIR-PORTWHITE PAPER*, 2019.
93. S. Kirkpatrick, C. D. Jr. Gelatt, and M. P. Vecchi. Optimisation by simulated annealing. *Science*, 220:671–680, 1983.
94. J. Kirton and E. Brooks. *'Cells in industry: Managing teams for profit'*, 1994.
95. F. Koenig, P. Found, and M. Kumar. Condition monitoring for airport baggage handling in the era of industry 4.0. *Journal of Quality in Maintenance Engineering*, 25, 02 2019.
96. S.H. Koopman. *A new approach to baggage handling at airports. Researching the effects of introducing flexible buffering by use of the 'pull concept*. PhD thesis, Delft University, 2018.
97. T.C. Koopmans and M.J. Beckmann. *Assignment problems and the location of economic activities*. Bobbs-Merrill reprint series in economics. Cowles Foundation for Research in Economics at Yale University, 1957.
98. S. Vatan Korkmaz, J. A. Hoyle, G. G. Knapik, R. E. Splittstoesser, G. Yang, D. R. Trippany, P. Lahoti, C. M. Sommerich, S. A. Lavender, and W. S. Marras. Baggage handling in an airplane cargo hold: An ergonomic intervention study. *International Journal of Industrial Ergonomics*, 36(4):301–312, 2006.
99. P. Kotler and G. Armstrong. *Principles of marketing*. Always learning. Pearson, 2014.
100. J. R. Koza. *Genetic programming: On the programming of computers by means of natural selection*. Cambridge, MA: MIT Press, 1992.
101. K. Kuhn and S. Loth. Airport service vehicle scheduling. 2009.
102. M. Laguna. *Tabu Search*, pages 741–758. Cham: Springer International Publishing, 2018.
103. D.P. Landau and K. Binder. *A guide to Monte Carlo simulations in statistical physics*. Cambridge University Press, 2005.
104. E. L. Lawler. *Combinatorial optimization: networks and matroids*. Courier Corporation, 2001.
105. A.M. Lee. *Applied queueing theory*. Studies in management. Macmillan, 1966.
106. Loo Hay Lee, Huei Chuen Huang, Chulung Lee, Ek Peng Chew, Wikrom Jaruphongsa, Yean Yik Yong, Zhe Liang, Chun How Leong, Yen Ping Tan, Kalyan Namburi, Ellis L. Johnson, and Jerry Banks. Simulation of airports/aviation systems: Discrete event simulation model for airline operations: Simair. In Stephen E. Chick, Paul J. Sanchez, David M. Ferrin, and Douglas J. Morrice, editors, *Winter Simulation Conference*, pages 1656–1662. ACM, 2003.
107. R.M.R. Lewis. *A guide to graph colouring: Algorithms and applications*. Berlin: Springer, 2016.
108. The Free Library. Dcvs are sexy again. 1999.
109. A. López-Carresi, B. Wisner, and M. Fordham. *Disaster management: International lessons in risk reduction, response and recovery*. Routledge, 2014.
110. M.-L. Lu, J. Dufour, E. Weston, and W. Marras. Effectiveness of a vacuum lifting system in reducing spinal load during airline baggage handling. *Applied Ergonomics*, 70:247–252, 2018.

111. L. G. Luther. Environmental impacts of airport operations, maintenance, and expansion. Environmental Science, Engineering, 2007.
112. D. Ma, J. Gausemeier, X. Fan, and M. Grafe. *Virtual reality & augmented reality in industry*. Springer Publishing Company, Incorporated, 2011.
113. K. Mainzer. *Thinking in complexity: The computational dynamics of matter, mind, and mankind*. Physics and astronomy online library. Springer Berlin Heidelberg, 2003.
114. M. Malek, M. Guruswamy, M. Pandya, and H. Owens. Serial and parallel simulated annealing and tabu search algorithms for the traveling salesman problem. *Annals of Operations Research*, 21(1):59–84, 1989.
115. Á. Marín. Airport management: taxi planning. *Annals OR*, 143(1):191–202, 2006.
116. B. Marr. What everyone must know about industry 4.0. *Forbes*, 2016.
117. S. Martello and P. Toth. *Knapsack problems: Algorithms and computer implementations*. Chichester: John Wiley & Sons, 1990.
118. V. Mayer-Schönberger and K. Cukier. Big data: A revolution that will transform how we live, work, and think. *Houghton Mifflin Harcourt*, 2013.
119. B. Mayoh, E. Tyugu, and J. Penjam. *Constraint programming*. Nato ASI Subseries F:. Springer Berlin Heidelberg, 2013.
120. D. Merkle and M. Middendorf. On solving permutation scheduling problems with ant colony optimization. *International Journal of Systems Science*, 36, 2004.
121. A.K. Mishra. *Particle swarm optimization: Bio inspired optimization technique*. LAP LAMBERT Academic Publishing, 2017.
122. A. Mortimer. Gatwick's bag factory. 1999.
123. M. Mutingi and C. Mbohwa. *Grouping genetic algorithms: Advances and applications*. Studies in Computational Intelligence. Springer International Publishing, 2016.
124. O. N. M. Lenior. Airport baggage handling - where do human factors fit in the challenges that airports put on a baggage system? *Work*, 41:5899–5904, 2012.
125. K. Nice. How baggage handling works. *science.howstuffworks.com*, 2001.
126. Y. Nikulin and A. Drexl. Theoretical aspects of multicriteria flight gate scheduling: deterministic and fuzzy models. *Journal of Scheduling*, 13(3):261–280, 2010.
127. J. R. Beasley, N. J. Ashford and P. Coutu. *Airport operations, 3rd Edition*. McGraw-Hill Education, 2012.
128. A. Nyberg, T. Westerlund, and A. Lundell. *Improved discrete reformulations for the quadratic assignment problem*, pages 193–203. Berlin, Heidelberg: Springer, 2013.
129. C. Oftring. An entry into the batch building concept for baggage handling systems, 2013.
130. C. Oftring. Improve baggage handling efficiency with multi-purpose baggage storage, 2013.
131. I. H. Osman and N. Christofides. Capacitated clustering problems by hybrid simulated annealing and tabu search. *International Transactions in Operational Research*, 1:317–336, 1994.
132. J. Pagani, A. O. Abd El Halim, Y. Hassan, and S. Easa. User-perceived level-of-service evaluation model for airport baggage-handling systems. *Transportation Research Record*, 1788(1):33–42, 2002.
133. G. Pahl, W. Beitz, and K. Wallace. *Engineering design: A systematic approach*. Design Council, 1996.
134. V. Pareto. Cours d'economie politique. *F. Rouge, Lausanne*, I and II, 1988.
135. M. de Zee Pascal Madeleine, A. Samani, and U. Kersting. *Biomechanical assessments in sports and ergonomics*. IntechOpen, Rijeka, 2011.

136. D. Patterson. *Introduction to artificial intelligence and expert systems*. Prentice-Hall, Inc., 1990.

137. R. Pikaar. Human factors engineering to reduce workload of baggage handling. pages 10–14, 2018.

138. A.D.J. Pinder, Health, and Safety Laboratory (Great Britain). *Benchmarking Manual Handling Assessment Charts (MAC)*. Health and Safety Laboratory, 2002.

139. M. L. Pinedo. *Scheduling: Theory, algorithms, and systems*. Springer Publishing Company, Incorporated, 3rd edition, 2008.

140. A. Polater. Managing airports in non-aviation related disasters: A systematic literature review. *International Journal of Disaster Risk Reduction*, 31:367–380, 2018.

141. J. T. Presby and M. L. Wolfson. An algorithm for solving job sequencing problems. *Management Science*, 13(8):B-454–B-464, 1967.

142. L. Prica. Big data in digital marketing. *medium.com*, 2019.

143. T. Pyzdek and P.A. Keller. The six sigma handbook. *McGraw-Hill Professional*, 2014.

144. Ingo. Rechenberg. *Evolutionsstrategie; optimierung technischer systeme nach prinzipien der biologischen evolution. Mit einem nachwort von Manfred Eigen*. Frommann-Holzboog [Stuttgart-Bad Cannstatt], 1973.

145. B. Rekiek and A. Delchambre. *Assembly Line Design: The Balancing of Mixed-Model Hybrid Assembly Lines with Genetic Algorithms*. 2006.

146. B. Rekiek, A. Delchambre, and H. Saleh. Handicapped person transportation: An application of the grouping genetic algorithm. *Engineering Applications of Artificial Intelligence*, 19:511–520, 2006.

147. A. Richter. *Gepäcklogistik auf Flughäfen: Grundlagen, Systeme, Konzepte und Perspektiven*. SpringerLink : Bücher. Springer Berlin Heidelberg, 2012.

148. B. Roy. Classement et choix en présence de points de vue multiples (la méthode electre). *Revue Française d'Informatique et de Recherche Opérationnelle*, 8:57–75, 1968.

149. R. Y. Rubinstein and D. P. Kroese. *Simulation and the Monte Carlo method*. Wiley Series in Probability and Statistics, 2 edition, 2007.

150. H. Saleh. *Artificial Intelligence and Geoinformation Technologies for Disaster Risk Reduction and Management*. Rabban Publishing LTD, 2016.

151. H. Saleh. An artificial intelligent design for gps surveying networks. *GPS Solutions*, 7:101–108, 2003.

152. H.A. Saleh. Artificial intelligence for optimizing the gnss carrier phase-based positioning. In *Proceedings of the 2003 National Technical Meeting of The Institute of Navigation*, pages 407–416, 2003.

153. R.K. Sarin. *Multi-attribute utility theory*, pages 526–529. New York, NY: Springer 2001.

154. S. C. Savvides. Risk analysis in investment appraisal. MPRA Paper 10035, University Library of Munich, Germany, 1994.

155. A. Schärlig. Décider sur plusieurs critères, panorama de l'aide à la décision multicritère. *Presses polytechniques et universitaires romandes*, 1985.

156. R. Scholing. Baggage handling: Achieving operational excellence. *International Airport Review*, 2014.

157. A. Schrijver. *Theory of linear and integer programming*. Wiley-Interscience series in discrete mathematics and optimization. John Wiley & Sons, Inc., 1999.

158. H.P. Schwefel. *Evolution and optimum seeking*. Sixth Generation Computer Technologies. Wiley, 1995.

159. P. Sen and J-B. Yang. Multiple criteria decision support in engineering design. *Springer-Verlag*, 1998.

160. Y. Shi and R.C. Eberhart. Empirical study of particle swarm optimization. *Evolutionary Computation*, 3:1950 Vol. 3, 1999.

161. A. Silven. *Autonomous transport robots in baggage handling systems*. Delft University of Technology, 2018.

162. H. A. Simon. The sciences of the artificial. 3rd edition, Cambridge, MA: The MIT Press, 1981.

163. P. Singhala, D. Shah, and B. Patel. Temperature control using fuzzy logic. *International Journal of Instrumentation and Control Systems*, 4, 2014.

164. SITA. *SITA baggage report 2017*. SITA, 2017.

165. Skelex. The ultimate exoskeleton for overhead work. 2020.

166. J.W. Smeltink, M. Soomer, P. de Waal, and R. Mei. An optimisation model for airport taxi scheduling. *Informs Journal on Computing - INFORMS*, 2004.

167. M. Sol and M. Savelsbergh. The general pickup and delivery problem. *Transportation Science*, 29:17–29, 1995.

168. H. Sousa, R. Teixeira, H. Lopes Cardoso, and E. Oliveira. Airline disruption management - dynamic aircraft scheduling with ant colony optimization. *ICAART 2015 - 7th International Conference on Agents and Artificial Intelligence, Proceedings*, 2:398–405, 2015.

169. D. Sriram, R. Banares-Alcantara, V. Venkatasubramanian, A. Westerberg, and M. Rychener. Knowledge-based expert systems: an emerging technology for cajd in chemical engineering. *Carnegie Mellon University - Research Showcase CMU*, DRC-06-76-84, 1984.

170. STAC. *Détermination de la capacité d'un aéroport*. dgac - France, 2005.

171. STAC. *Capacité des aérogares passagers - Guide technique*. dgac - France, 2010.

172. STAC. *La capacité aéroportuaire - Guide technique*. dgac - France, 2018.

173. N. Stanton, W.-C. Li, and D. Harris. Ergonomics and human factors in aviation. *Ergonomics*, 60:150–150, 2017.

174. P. M. Swamidass, editor. *ABC analysis or ABC classification*, pages 1–2. Boston, MA: Springer, 2000.

175. D. Symonds. Vehicle for change. Annual showcase 2022 2021.

176. E. Taillard. Some efficient heuristic methods for the flow shop sequencing problem. *European Journal of Operational Research*, 47(1):65–74, 1990.

177. A.N. Tarău, B. De Schutter, and J. Hellendoorn. Predictive control for baggage handling systems using mixed integer linear programming. In *Proceedings of the 5th IFAC International Conference on Management and Control of Production Logistics (MCPL 2010)*, pages 16–21, Coimbra, Portugal, 2010.

178. R. K. Taylor and D. L. Anderson. Early bag storage system. Patent wo9510467 (a1) mi usa, JERVIS B. WEBB COMPANY, 1994.

179. P. Toth and D. Vigo. *Vehicle routing: Problems, methods, and applications, 2nd Edition*. Philadelphia, PA: Society for Industrial and Applied Mathematics, 2014.

180. U. Lykkegaard and D. Fransen. Loading of aircraft baggage containers. Patent wo2015/197070 al denmark, Crisplant, 2015.

181. K. P. Valavanis, A. I. Kokkinaki, and S. G. Tzafestas. Knowledge-based (expert) systems in engineering applications: A survey. *Journal of Intelligent and Robotic Systems*, 10(2):113–145, 1994.

182. L.L.P. van Rijen. *State of the art survey of baggage handling systems control and automated equipment.* Delft University of Technology, 26-01-2016.

183. Vanderlande. Fleet: all systems go at rotterdam. Marketing paper, Vanderlande, 2021.

184. D. Vanderpooten. The interactive approach in mcda: A technical framework and some basic conceptions. *Mathematical and Computer Modelling,* 12(10):1213–1220, 1989.

185. Ph. Vincke. *Multicriteria decision-aid.* New York, NY: John Wiley & Sons, 1992.

186. J. Vom Brocke and M. Rosemann. *Handbook on business process management 1 - Introduction, methods, and information systems.* Springer-Verlag, 2010.

187. J. Vom Brocke and M. Rosemann. *Handbook on business process management 2 - Strategic alignment, governance, people and culture.* Springer-Verlag, 2010.

188. S. Wang, J. Wan, D. Zhang, D. Li, and C. Zhang. Towards smart factory for industry 4.0: a self-organized multi-agent system with big data based feedback and coordination. *Computer Networks,* 101:158–168, 2016.

189. M. Weske. Business process management: Concepts, languages, architectures. *Springer-Verlag,* 2012.

190. M. Widmer and A. Hertz. A new heuristic method for the flow shop sequencing problem. *European Journal of Operational Research,* 41(2):186–193, 1989.

191. A. Willihnganz. Software that writes software: Genetic programming is the new frontier: A human creates the environment, and a computer hacks the code. *salon.com,* 1999.

192. J. Wiltshire. Airport competition: Reality or myth? *Journal of Air Transport Management,* 67(C):241–248, 2018.

193. L.A. Wolsey. Mixed integer programming. In *Wiley Encyclopedia of Computer Science and Engineering.* John Wiley & Sons, Inc., 2008.

194. J.P. Womack and D.T. Jones. *Lean Thinking: Banish Waste and Create Wealth in Your Corporation.* Simon & Schuster, 2013.

195. R. Xinhui. The application of six sigma methodology for check-in service in airport. In *ICSSSM11,* pages 1–5, 2011.

196. J. Xu and G. Bailey. The airport gate assignment problem: Mathematical model and a tabu search algorithm. *Proceedings of the 34th Hawaii International Conference on System Sciences,* 3, 2001.

197. S. Young and A.T. Wells. *Airport planning and management 6/E.* McGraw-Hill Education, 2011.

198. L. Zadeh. Fuzzy sets. *Information and Control,* 8:338–353, 1965.

199. Y. Zeinaly, B. De Schutter, and H. Hellendoorn. An integrated model predictive scheme for baggage-handling systems: Routing, line balancing, and empty-cart management. *IEEE Transactions on Control Systems Technology,* 23(4):1536–1545, 2015.

Index

Printed in the United States
by Baker & Taylor Publisher Services

Printed in the United States
by Baker & Taylor Publisher Services